American Institute of Chemical Engineers

Topical Conference on Applications of Micro-Reactor Engineering 2007

Held at the 2007 AIChE Spring National Meeting

April 22-27, 2007
Houston, Texas, USA

Printed from e-media with permission by:

Curran Associates, Inc.
57 Morehouse Lane
Red Hook, NY 12571
www.proceedings.com

ISBN: 978-1-60423-350-6

Some format issues inherent in the e-media version may also appear in this print version.

ISBN: 978-1-60423-350-6

American Institute of Chemical Engineers
Proceedings
Three Park Avenue
New York, NY 10016-5991
Phone: 212-591-8100

www.aiche.org

American Institute of Chemical Engineers

Topical Conference on Applications of Micro-Reactor Engineering
2007

TABLE OF CONTENTS

Session 7 – Exergy, Cost and Eco-Efficiency Analysis in Micro-Systems

Session 8 – Fine Chemicals and Pharmaceuticals Production in Micro-Systems Part I

Session 32 – Fine Chemicals and Pharmaceuticals Production in Micro-Systems Part II

Session 35 – Polymerization in Micro-Reactors

Session 49 – Commercialization of Micro Process Technology-Panel Discussion

Session 73 – Fluid Manipulation/Nanofluidics

Session 74 – Mixing and Heat Exchange in Micro-Reactors

Session 89 – Emulsions and Dispersions

Session 95 – Micro Technology in Catalyzed Reaction Systems – Part I

Session 113 – Micro Technology in Catalyzed Reaction Systems – Part II

Session 121- Particles and Beads

Exergetic Efficient Distillation in Microchannel Architecture

Laura Silva[1], Daniel Weidert[2], Robert Litt[3], Anna Lee Tonkovich[4], Ravi Arora[3], and Maddalena Fanelli[3]. (1) Business Development, Velocys, Inc., 7950 Corporate Boulevard, Plain City, OH 43064, (2) Project Management, Velocys, Inc., 7950 Corporate Boulevard, Plain City, OH 43064, (3) Velocys, Inc., 7950 Corporate Blvd., Plain City, OH 43064, (4) Technology Development, Velocys, Inc., 7950 Corporate Boulevard, Plain City, OH 43064

The technology of separating liquid-liquid mixtures via distillation has been practiced for literally thousands of years, yet it remains a highly energy-intensive, although ubiquitous unit operation for performing chemical separations. Modern distillation consumes over 5 quadrillion BTUs annually, in the U.S. alone. Over the past several decades, new distillation column, tray, and packing designs have yielded incrementally improved efficiency, while larger improvements have remained elusive. Microchannel architecture permits a radically new approach to distillation; one that has the potential to provide a step change improvement in energy consumption as well as lower capital costs and other process advantages.

Exergy is the maximum theoretical amount of work that can be extracted from a physical system by exchanging matter and energy with large reservoirs in a reference state. The exergetic efficiency of a process is an objective performance measure of a real system as compared to a reversible system with no lost work. While energy is conserved, exergy can be destroyed. Using this definition and understanding that distillation consists of heat transfer for vaporizing and condensing, and mass transfer for the separation of the mixture; a distillation unit can be described as an exergy converter where thermal exergy is converted into chemical exergy. Exergetic inefficiencies in conventional distillation result from both technological and capital cost constraints. For example, increased efficiency in conventional distillation towers can be gained by incorporating side reboilers at multiple stages, which utilize waste heat mediums at temperatures lower than the bottom reboiler heat medium. However, the capital cost of adding multiple reboilers and any associated equipment to conventional distillation towers can make this option cost prohibitive. During times of cheap energy, this trade-off between energy cost and capital cost almost always favors the solution with lower capital cost.

Microchannel distillation technology enhances mass transfer by contacting thin vapor and liquid films in very small channels, dramatically reducing the Height-to-an-Equivalent Theoretical Plate (HETP), from more than 12 inches to less than 2 inches. Microchannel architecture also allows temperature profiles to be tightly controlled along the length of the channels using adjacent heat exchange channels integrated in the separation hardware. Enhancements in both mass transfer and heat integration enable dramatically improved performance and new process flow sheet options which were previously seen as commercially unattractive. Microchannel technology is scaled to commercial plant capacities by increasing the numbers of parallel channels conducting identical processes. Techniques for design and fabrication of scaled-up microchannel processing hardware have been developed by Velocys for a range of large-scale micro reactor applications.

One commercially important distillation application is the C2 splitter, which separates ethane from ethylene. In this application alone, the microchannel distillation process could reduce energy consumption by up to 20% for the entire process, which translates to over 40 trillion BTU/year by 2020, in the U.S. alone. This presentation will cover the technical approach, benefits, and recent accomplishments in the development of the microchannel distillation process, with a specific focus on the C2 splitter.

High Energy Efficiency Micro Channel Raector for the Intermediate Temperature Fuel Cell Systems

Takashi Shimazu[1], Hiroshi Aoki[2], Tomohisa Wakasugi[1], Hiroyuki Mitsui[1], Takatoshi Masui[3], and Kenji Kimura[4]. (1) Powertrain Thermofluid Technologies LAB., Toyota Central Research & Development Laboratories., INC., 411 Nagakute, AICHI-GUN, 480-1192, Japan, (2) Development Div., Toyota Central Research & Development Laboratories., INC., 411 Nagakute, AICHI-GUN, 480-1192, Japan, (3) Fuel Cell Development Div., Toyota Motor Corporation Higashifuji Technical Center, 1200 Mishuku Susono, Shizuoka, 410-1193, Japan, (4) Fuel Cell Development Center, Toyota Motor Corporation, 1, TOYOTA-CHO TOYOTA, AICHI, 471-8572, Japan

Toyota introduced the world's first Fuel Cell Hybrid Vehicle, the FCHV, into the market in 2002. Although fuel cell vehicles have great potential as future vehicles, it is also realized that further breakthrough technologies to solve essential problems, such as cost, energy efficiency and cruising range, are needed for Fuel Cell vehicles to be competitive with Internal combustion engine, the ICE vehicles. As a means of solution to the problems, Toyota proposed the high efficiency fuel cell system consisted of a combination of intermediate temperature (300-400 degrees-C) hydrogen membrane fuel cell, HMFC, with a heat exchanged hydrogen generator in 2004.The use of an on-board reformer with a Fuel Cell system is a very attractive option for FC vehicles because of its long cruising range and no need for a new hydrogen infrastructure. Two characteristics are essential for an on-board Fuel Cell used with a reformer, a high power density and an intermediate operating temperature. However, the on-board reformer for HMFC system requires the processing of a liquid hydrocarbon fuel to produce a hydrogen-rich gas stream suitable for consumption by the fuel cell. Conventional technology for hydrocarbon steam reforming experiences heat transfer limitations resulting in long residence times and large equipment. As a result, most automotive on-board reforming efforts have targeted partial oxidation (POX) and autothermal (ATR) reforming approaches that the heat by injecting air along with the reactants as opposed to heat exchanged reactor. However, heat exchanged reforming offers several potential advantages over the POX or ATR approaches, including the fact that steam reforming can combust waste anode gas as fuel to provide the necessary heat input, allowing it to be more efficient. POX and ATR, which are thermally neutral or exothermic, cannot use the waste anode gas in this way. Several methods have been proposed to achieve this system. However both conversion rate and reformer efficiency are not adequate. In this paper, we introduce a compact and high energy efficiency heat exchanged micro channel reactor which generates a supply of hydrogen-rich gas to an intermediate temperature hydrogen membrane fuel cell, which we call HMFC. In addition, we introduce the catalyst coating technique appropriate to the metallic substrate of a micro channel reactor. One of the main problems in the use of microreactors for heterogeneous catalytic reactions is the development of catalytically active microporous support materials. Several solutions have been proposed to overcome this problem, such as filling the channels with the catalyst, or scale down of industrial reactors, or anodic oxidation of microstructured aluminium foils. However all solutions are not adequate from the viewpoint of trade-off between adhesive and film thickness. In this paper, we will introduce the catalyst coating technique appropriate to the metallic substrate of micro channel reactors. In this paper, we will introduce two new leads for future FC vehicles. The first is a compact and high energy efficiency heat exchanged micro channel reactor. The second is a catalyst coating technique. As a result of the two techniques, the overall core volume of the reactor is 33.7 cc to generate 4.5 kW LHV hydrogen. While reforming iso-octane at a space velocity 50,000 h-1, the reformer efficiency achieved more than 80% by recovery of heat and steam after integration with an intermediate temperature fuel cell. In the catalyst coating technique, immobilization of nano particles prepared from a colloid solution and metal salt solution produces good adhesion of catalyst metal substrate to the micro channel reactor. The catalyst coats were best suited to coat micro channel reactor due to ultra-thin layers ranging in thickness from 0.1 to 50 micrometers. The present reformer system has successfully demonstrated that the high efficiency reactor can achieve a long cruising range. Furthermore, the current demonstration is a major step towards the goal of future FC vehicles.

Silicon Carbide Microreactors with Catalyst Carrier Coatings

Frank A. Meschke, ESK Ceramics GmbH & Co. KG, Max-Schaidhauf-Strasse 25, D-87474 Kempten, Germany

Micro reaction technology has become very popular with chemical engineers over the past decade due to the enhanced safety features, modular architecture, flexibility, improved performance and other benefits it offers. Hermetically gas-tight micro reactors made from sintered silicon carbide ceramics complement the metal, glass and polymer systems currently available. SiC micro reactors are characterized by excellent high temperature stability and almost complete resistance to both corrosion and erosion. Furthermore, SiC-micro reactors can be manufactured as a ceramic monolithic structure for almost any internal geometry desired. High heat transfer rates, the absence of leakage and stability at high pressures exceeding 10 bars and at temperatures well above 800°C are key features.

The aim of this presentation is to demonstrate available design possibilities and applications of monolithic SiC microreactors and variants with catalytic coatings.

Microreaction Technology and Its Route to Production Plant

Barbara Pieters, Geraldine Andrieux, and Jean-Christophe Eloy. Yole Développement, 45 rue Sainte-Genevieve, Lyon, F-69006, France

Microreaction technology has proven its added value in the R&D department of large chemical companies for the synthesis of new molecules and the development of new reaction pathways. This market remains a low volume market today (about 10 M € for MRT modules and systems). The stake for the development of microreaction market lies in the transfer of the technology from lab use to large scale chemical productions.

This paper will thus be dedicated to the presentation of the expected implementation of microreaction technologies for production purpose. We will higlight the advantages perceived for MRT both from a technological and economical point of view. We will also illustrate remaining hurdles.

Our analysis will be illustrated by specific case studies of MRT early adopters in the chemical industry. We will present how large chemical companies are working with microreactors and what evolution occurred in the past 5 years.

This new trend has led to an evolution in the MRT market structuration. Several companies have entered the field recently with a new business model. They are not providing microstructured parts but rather enabling technologies for a better and more efficient use of microreactors. We will thus present players and products designed for large scale production, as well as companies whose business model is to support and engineer microreaction plants for large scale production.

As a conclusion, we will present the expected future evolution of the MRT field.

Microreactor Technology as an Integrated Method for Producing Fine Chemicals

Dominique M. Roberge, Exclusive Synthesis, Process R&D, Lonza Ltd., Lonzastr., Visp, Switzerland

Microreactor Technology enables processes to be run in a continuous manner using a minimal quantity of reagent. Thus, it permits the rapid and scalable development of continuous processes in the fine chemical and pharmaceutical industries. Under such circumstances, advantages are associated in a first instance from the continuous way of operation and in a second instance to the micro structure itself such as the good thermal control. It is important to differentiate from where the advantage is coming from because it will significantly influence the scale up strategy for larger productions.

In the fine chemical industry, productions are made in multi-purpose plants of high flexibility. The integration of a continuous system is such an environment is feasible given appropriate modules are employed with (i) good chemical resistance (i.e. glass, Hastelloy, Teflon), with (ii) excellent material stability over a large range of temperatures, and with (iii) ease of connection avoiding dead volume. In addition the various modules need to take into account the physico-chemical properties of the reaction such as the reaction kinetics („± residence time) and the reaction phases (solid - liquid - gas). This approach leads systematically to a Toolbox concept.

A detailed analysis at Lonza [1] showed that ca. 50% of the reactions studied could fit into a microreactor based on their kinetic. However, by taking into account the reaction phases, this number shrinks to ca. 20% of potential candidates because a solid phase is present in more than 60% of the cases. In addition, the reactions were classified into 3 classes namely Type A (mixing controlled reactions), Type B (rapid but kinetically controlled reactions), and Type C reactions (batch reactions with thermal hazard).

The goal of Lonza in the past 3 years was to define the needs of this Toolbox. Collaboration with partners such as Corning, Siemens, and IMM were undertaken. We developed 3 reactor modules now called: Module A, B, and C. The presentation will detail the 3 type of modules. These modules are now commercially used at Lonza and several tons of products were synthesized with them. Some manufacturing examples will be presented showing the applicability of Microreactor Technology as an integrated method for producing fine chemicals.

In addition, the lecture will also address in a detailed manner the scale-up strategies available for larger productions. The requirements for a process in clinical phases and later in commercial production are quite different and this difference drives the distinctive needs in the supply chain of fine chemicals. These factors strongly influence the scale-up, multi-scale, or numbering-up strategies.

Modular Microchemical Engineering - a Tool for Process Intensification

Olaf Stange[1], Alexander Azzawi[1], Carmine Raffa[1], Frank Schael[1], Hans-Erich Gasche[2], Marcus Grünewald[2], Shaibal Roy[3], and Eric Boonstra[3]. (1) Ehrfeld Mikrotechnik BTS GmbH, Mikroforum Ring 1, 55234 Wendelsheim, Germany, (2) Bayer Technology Services GmbH, Leverkusen, 51368, Germany, (3) Bayer Corporate and Business Services, 8500 West Bay Road, Baytown, TX 77520

Microchemical engineering opens new pathways for the development of economical and innovative processes and intensified processes. In the chemical and pharmaceutical industry, this technology plays an increasingly important role in research, development and production because of its decisive advantages over conventional technology.

With the Modular Microreaction System, a powerful development and production tool combining the advantages of microprocess technology with great flexibility and a wide scope of application for process development and intensification will be shown. The system comprises individual microstructure-equipped modules that can be freely combined. These modules perform basic process operations such as mixing, dispersing, heat transfer, etc. The individual microreaction modules can be freely arranged on a solid base plate, enabling the fast and flexible setup of a wide variety of microreaction plants.

The Modular Microreaction System can be used for liquid phase, gas phase and multiphase reactions in heterogeneous gas-liquid systems. Operating parameters include a temperature range of –20 to 200 °C; operating pressures up to 100 bar (at room temperature); and rated throughputs of up to 25 L/h for low-viscosity media. Special modules available significantly extend this process range. Among the items available are special reactors for reaction temperatures from –100 °C to 500 °C and stand-alone, high-performance production modules with rated throughputs of several thousand liter per hour.

The Modular Microreaction System provides the user with a flexible, safe and cost-effective platform for the development of new synthesis methods for the production of chemical compounds and improving existing processes through intensification. Continuous operation and the more efficient process control made possible by microprocess technology often result in significantly shorter development times and thus reduced costs.

Examples for applications in lab scale as well as in pilot scale will be shown for different reaction systems. In addition new microreaction modules will be presented and authors will discuss the advantages of modular plant concepts for process intensification.

Web Page: www.ehrfeld.com

Scale-out Concept for Falling Film Micro Reactors

Ulrich Krtschil, Volker Hessel, Patrick Löb, Bhanu Kiran Vankayala
Institut für Mikrotechnik Mainz GmbH, Mainz, Germany

Over the past years the standard falling film microreactor (FFMR) developed by our institute [1] was successfully used for numerous gas-liquid reactions in the laboratory scale [2-5]. In the context of the German public funded research project µ.Pro.Chem [6] two types of enlarged falling film microreactors were developed following different scale-out concepts for the pilot and production scale (see Figure 1). These concepts differ in the principles of construction and in the extent of internal numbering-up.

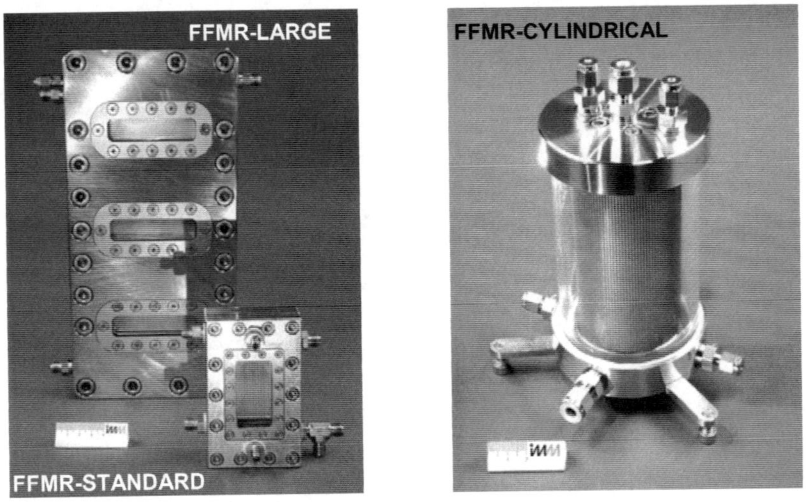

Figure 1. *FFMR- LARGE and FFMR-CYLINDRICAL compared with FFMR- STANDARD*

The new developed FFMR-CYLINDRICAL was investigated using the same test reaction as used for the IMM standard FFMR, an oxidation reaction. Comparable yields could be reached at tenfold higher flow rate (see Figure 2).

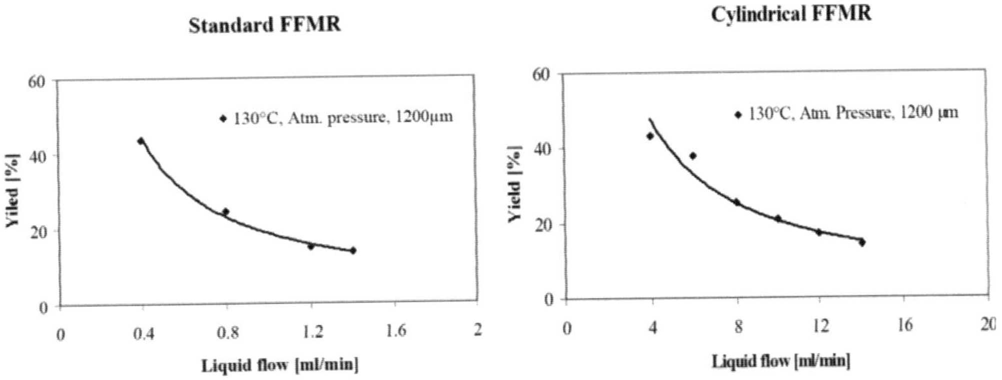

Figure 2. *Comparison of the yields for an oxidation reaction [8]*

In addition executed experiments resulted both in a good liquid equal distribution (mean deviation 7%) and also in a good gas equal distribution, since in a decoloration reaction experiment no bleached areas were visible. This could be reached by an advanced distribution system. In one set of experiments the oxidation of an organic compound was performed with a maximum conversion of 80%. This reaction is in analogy to a real-case industrial process undergone in parallel. The predictability of scale-out by increase in channels dimensions was tested using the Nusselt and the more advanced Kapitza and Feind correlations for calculating the film thickness. These equations predict same liquid film thickness for different channel dimensions for a given flow rate since the wetted area is assumed to be the same. However, experiments show varying conversion, which probably is due to an insufficiency of the mentioned correlations which do not take into account surface and wall effects and due to varying film thickness (see Figure 3).

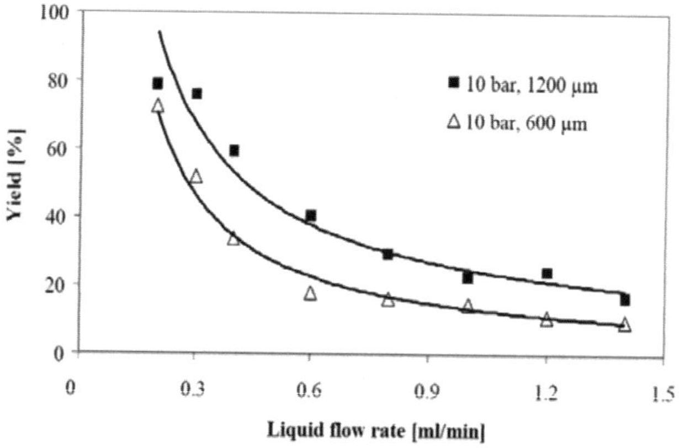

Figure 3. *Yield in oxidation reaction as function of liquid flow rate for two different plates [8]*

The latter is corroborated by experimental determination of the film thickness done by confocal microscopy recently [7]. In addition to investigations in our institute first tests with a pilot scale FFMR, reaching liquid throughputs up to 1 l/h, were successfully carried out in industry based on the results obtained with the Standard FFMR. The aim of the scale-out concept is to transfer the industrial used gas-liquid reaction from the lab scale to the production scale. In future these more experimental works will be supplemented by modeling and by correlations obtained from dimensional analysis.

References:

1. Wille, Ch., Dissertation, TU Clausthal, (2000).
2. Jähnisch, K., Baerns, M., Hessel, V., Ehrfeld, W., Haverkamp, V., Löwe, H., Wille, C., Guber, A.,.Direct fluorination of toluene using elemental fluorine in gas/liquid microreactors. Journal of Fluorine Chemistry (2000), 105, 117–128.
3. Löb, P., Löwe, H., Hessel, V., Fluorinations, Chlorinations and Brominations of Organic compounds in Micro Reactors, Journal of Fluorine Chemistry (2004), 125 (11) 1677-1694.
4. Zanfir, M., Gavriilidis, Wille, Ch., Hessel, V., Carbon Dioxide Absorption in a Falling Film Microstructured Reactor: Experiments and Modelling, Industrial Engineering Chem. Res. (2005), 44 1742-1752.
5. Jähnisch, K. et al., IMRET-9, 2006, Abstracts: "Ozonation of Olefins in Micro-structured Reactors" and "Multistep process for synthesis of pharmaceutical intermediates by ozonation and hydrogenation using micro chemical engineering with integrated fiber-optical diamond ATR sensor".
6. BMBF project µ.Pro.Chem http://www.mstonline.de/foerderung/projektliste/detail_html?vb_nr=V3MVT021.

7. Yeong, K-K., Gavrillidis, A., Zapf, R., Kost, H-J., Hessel, V., Boyde, A., Characterisation of liquid film in a microstructed falling film reactor using laser scanning confocal microscopy, Experimental Thermal and fluid science 30 (2006) 463-472.
8. Vankayala, B. K., Löb, P., Hessel, V., Menges, G., Hofmann, Ch., Metzke, D., Kost, H.-J., Krtschil, U., Scale-up of Process Intensifying Falling Film Microreactors to Pilot Production Scale, International Journal of Chemical Reactor Engineering (2007) (submitted).

Application of Micro-Reactor Technology in Process Development

Xini Zhang, Chemical & Pharmaceutical Development, Johnson & Johnson Pharmaceutical Research & Development, L.L.C., 1000 Rt. 202, Raritan, NJ 08869

The microreactor technology is an efficient tool for kilogram scale syntheses in continuous mode and is particularly effective for hazardous reactions that do not allow scale up in conventional reactors. Applications to several classes of reactions including highly exothermic reactions, high temperature reactions, reactions with unstable intermediates and reactions involving hazardous reagents are discussed.

Kinetic Study of Catalytic Hydrogenation of Aromatic Nitro Compound in a Microchannel Reactor

Sunitha Tadepalli[1], Adeniyi Lawal[2], Jale Muslehiddinoglu[3], and Donald Kientzler[3]. (1) Dept. of Chemical, Biomedical and Materials Engineering, Stevens Institute of Technology, New Jersey Center for Microchemical Systems, Castle Point, Hoboken, NJ 07030, (2) New Jersey Center for Microchemical Systems, Dept. of Chemical, Biomedical and Materials engineering, Stevens Institute of Technology, Castle Point, Hoboken, NJ 07030, (3) Process R&D, Bristol-Myers Squibb Company, One Squibb Drive, New Brunswick, NJ 08903

Hydrogenation reactions are ubiquitous in the fine chemicals and pharmaceutical industries. The kinetics of fast three-phase hydrogenation reactions in conventional slurry or batch reactors is often limited by mass transfer of hydrogen gas through the liquid due to the limited solubility of hydrogen in organic substrates and solvents. The use of a microchannel reactor for such reactions provides improved mass transfer rates which may ensure that the reaction operates close to intrinsic kinetics. In the present study, hydrogenation of aromatic nitro compound was conducted in a packed-bed microreactor as a model hydrogenation reaction of importance to the pharmaceutical industry with the aim of investigating the kinetics of the reaction. Internal and external mass and heat transfer limitations were examined. Kinetic experiments were conducted in a differential reactor mode keeping the conversion of the reactant at less than 10%. Different rate laws using different mechanisms from the literature were considered to fit the experimental data. Experiments were also carried out in an integral reactor, and the integral reactor data were compared with the predictions of the theoretical models.

The Synthesis of Isotopically Labelled Compounds in Micro Reactors

P. Watts, Julian Hooper, Ben Wahab, and Charlotte Wiles. University of Hull, Hull, United Kingdom

Radiolabelled molecules are of growing importance within the pharmaceutical and biomedical fields since they may be used within a diverse range of applications. For instance, radiolabelled isotopomers of drug candidates and significant metabolites are used in drug metabolism studies within the pharmaceutical industry, these would most commonly use [14C] but may alternatively use other isotopes, furthermore tritium labelled isotopomers are routinely used in mode of action studies. A second very topical application of radiochemistry is in positron emission tomography (PET) studies where molecules are labelled with positron-emitting isotopes, which possess very short half-lives, such as carbon-11 (11C, t1/2 = 20.3 min) and fluorine-18 (18F, t1/2 = 109.7 min). PET is an increasingly important medical imaging technique used for research and clinical applications in the detection of cancer, coronary artery disease and brain disease.

If we consider the present situation within a radioisotope group, the first synthesis would be the 'warm' radiosynthesis in which the reaction is spiked with a small amount of radioactivity in order to provide information on overall radioaccountability. Once this trial is successfully completed, a full radiosynthesis ('hot') is undertaken to provide radiolabelled product, which would be repeated as required. The downside of this approach is that the 'warm' reactions can never be considered as truly predictive of the undiluted 'hot' reaction, even if they are carried out on the same chemical scale. Thus, for a radiochemical synthesis the scale out of micro reactors to provide the required quantity would certainly be far less of an issue than that originally perceived with batch reactions. Hence a trial reaction using undiluted radiochemicals could be carried out effectively and safely in a single micro reactor with minimal sample quantity. It is envisaged that once a process has been successfully carried out within a micro reactor and scaled up to the required quantity using a parallel approach, then this customised system will be retained specifically for this reaction sequence. In this manner the reproducibility of the synthesis is guaranteed and the variability otherwise introduced by the work being carried out by different chemists is removed.

In this paper we report that micro reactors may be used to prepare labelled compounds by conducting all optimisation experiments using unlabelled precursors and simply substituting the labelled derivatives once the optimisation is complete. A range of reaction chemistries will be outlined including PET and stable isotope derivatives.

Catalytic Hydrogenation of Aromatic Nitro Compound in a Micro-Reactor

Adeniyi Lawal[1], Sunitha Tadepalli[2], Jale Muslehiddinoglu[3], and Donald Kientzler[3]. (1) New Jersey Center for Microchemical Systems, Dept. of Chemical, Biomedical and Materials engineering, Stevens Institute of Technology, Castle Point, Hoboken, NJ 07030, (2) Dept. of Chemical, Biomedical and Materials Engineering, Stevens Institute of Technology, New Jersey Center for Microchemical Systems, Castle Point, Hoboken, NJ 07030, (3) Process R&D, Bristol-Myers Squibb Company, One Squibb Drive, New Brunswick, NJ 08903

In this work, a three phase solid catalyzed hydrogenation reaction of aromatic nitro compound was studied in a microreactor as a model pharmaceutical reaction. In the pharmaceutical industry, hydrogenation reactions constitute about 15-20% of the reactions and are generally carried out in batch reactors. Hydrogenation of nitro compounds falls under the class of reactions with fast kinetics on noble-metals, and is usually limited by external mass transfer in batch reactors. Besides, these reactions are highly exothermic in nature which necessitates effective heat removal. Ineffective mass transfer and non-uniform temperature distribution in the batch reactor can have deleterious effects on reactor performance including the formation of undesired side products. The use of microreactor for such reactions provides improved mass and heat transfer rates which may ensure that the reaction is close to intrinsic kinetics.

The main objective of this study is to compare the performance of the microreactor with a batch reactor using this model reaction. For this purpose, experiments were conducted in a packed bed micro-reactor and also in a semi-batch reactor. Internal and external mass transfer limitations were examined and kinetic experiments conducted in both reactor systems. Different rate laws using different mechanisms from the literature were considered to fit the experimental data. The mass transfer coefficient in the microreactor was found to be orders of magnitude greater than that in the semi-batch reactor. Furthermore, the effects of various operating conditions on the performance of the two reactor systems were investigated, and their performance measured by the space-time yield.

Synthesis of Au Nanorings in the Soft Template of SDS-PEG Cluster under Mild Conditions

Chunrong Wang YunFang[*] JieWen FENG

(School of Chemical and Material Engineering, Southern Yangtze University, Wuxi 214122, China)

Abstract

A novel method is reported to prepare Au nanorings from $HAuCl_4$ self-reduced by polyethylene glycol (PEG) in soft template composed of sodium dodecyl sulfate (SDS)-PEG 20000 cluster at ambient temperature and atmosphere, in which no hard template and extra reducer is needed. Transmission electron microscopy (TEM) image displays that the diameter of Au nanorings is 500 ± 50 nm. Electron diffraction (ED) pattern and XRD pattern show that the synthesized nanoring is monocrystal Au and grows preferentially along (111) lattice planes. UV-vis spectrum reveals that there is relatively stronger absorption band beyond 800 nm, which is attributed to big and anisotropic nanostructrues. The process of nanoparticles growing from Au nanospheres to nanoplates and eventually to nanorings is tracked by both TEM and SEM, which indicates that only the reduction period is monitored can the preferred Au nanostructures be obtained.

Keywords Au nanoring; soft template; cluster; SDS; PEG

1. Introduction

Metal nanoparticles have many important applications in many aspects such as catalysis process, optics, medicine, etc. (Cui, 2001; Daniel, 2004) it is well known that the physical, optical, and chemical properties of nanoparticles depend on their size and shape greatly. Therefore, it is very essential to develop a simple and effective preparation method of metal nanoparticles for well-controlled sizes and shapes. In recent years, many Au nanostructures such as nanospheres (Jana, 2001), nanorods (Nikoobakht, 2003), nanowires (Pei, 2004), nanodisks (Simakin, 2001), multipods (Kuo, 2005), nanocubes (Huang, 2006) have been synthesized. However, compared with the various morphological controls of Au nanostructures, the production of nanoplates and nanorings have met only few limited success (Wang, 2005; Guo, 2006; Li, 2006) (Metraux, 2003; Sun, 2003; Jiang, 2004). For example, so far the method for synthesis of Au nanorings is only using silver nanoplates as hard templates via a displacement reaction (Metraux and Cao et al., 2003; Sun and Xia, 2003; Jiang and Xu et al., 2004). The method is complicated and easy to pollute environment. We demonstrate herein a simple chemical route to synthesize Au nanorings. In the novel process, no metal hard template or extra reducer was needed and $H[AuCl_4]$ was self-reduced by PEG component in the soft template composed of SDS-PEG clusters (TSP) to form nanorings

2. Experimental

2.1 Chemicals

Polyethyleneglycol 20000(PEG 20000, chromatographic grade) as well as

[*] To whom correspondence should be addressed.

Tel. and Fax: 86-510-5865424. E-mail: yunfang@126.com

$HAuCl_4 \cdot 4H_2O$(analytical reagent) were from Sinopharm Chemical Reagent Co., Ltd. SDS (99.0%, Acros Organics) were from Aldrich. Deionized water(conductivity 7.8×10^{-7} S/cm).

2.2. Preparation of Au nanorings

In a typical experiment, an aqueous solution of 5 mM $H[AuCl_4]$ (1 mL) was added to the solution (19 mL) containing PEG 20000 and SDS. The volume was adjusted to 25 mL with water while stirring; the final concentrations were 0.2 mM $H[AuCl_4]$, 5 mM PEG 20000 and 3 mM SDS. Experiments were conducted at room temperature (25±0.5 °C) and without ambient light to avoid photoreactions of the Au complex. Yellow precipitate was observed 10 days later, which was collected by centrifugation, washed several times with water, and then resuspended in water. The resulting suspension was used for further characterization.

2.3. Characterization of Au nanoparticles

The samples for TEM and scanning electron microscopy (SEM) investigations were prepared by placing small drops of the suspension on carbon-coated copper grids and copper substrate, respectively. The samples were then allowed to dry at room temperature in a fume hood. The TEM images and electron diffraction patterns were taken using a Hitachi H-600 electron microscope operated at 100 kV. SEM images were obtained using FEI Quanta-200 operated at 15 kV. XRD was carried out with D/Max 2500 X-ray diffractometer using Cu (40 kV, 200 mA) radiation and the samples were prepared as thin layers on glass slides. The UV/Vis spectra were recorded on Purkinje General TU-1901 spectrophotometers.

3. Results and Discussion

The crystalline nature of the nanoring is shown in the X-ray diffraction (XRD) pattern (Figure 2). The XRD pattern shows sharp peaks corresponding to the (111), (200), (220) and (311) diffraction peaks of metallic Au, and thus indicates that the particles is composed of pure crystalline Au (Maye, 2000). Nate that the ratio of the intensity between the (200) and (111) diffraction peaks is about 0.11 much lower than the standard file (JCPDS: 0.52). These observations confirm that our nanoplates are primarily dominated by (111) factes (Kuo, 2005; Sun, 2005).

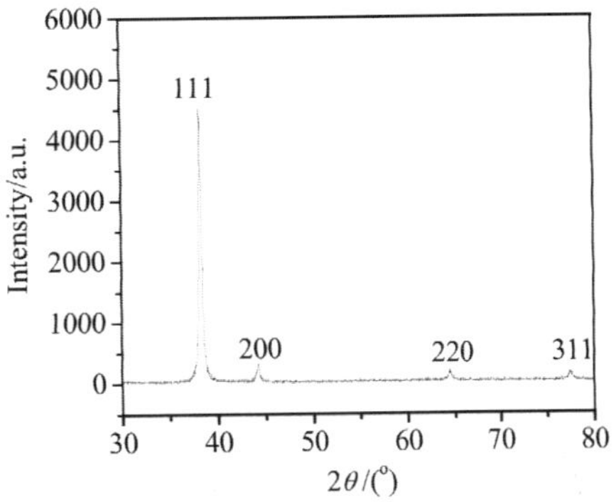

Figure 1 XRD pattern of the Au nanorings prepared from $HAuCl_4$ (0.2 mmol/L)-SDS (3 mmol/L)-PEG 20000 (5 mmol/L) system after 10 days.

The morphology of the nanoring was characterized by TEM (Figure 2). The TEM image shows that the diameter of Au nanoring is about 500±50 nm, and the ED pattern (inset of Figure 2) indicates that the Au nanoring is single crystalline.

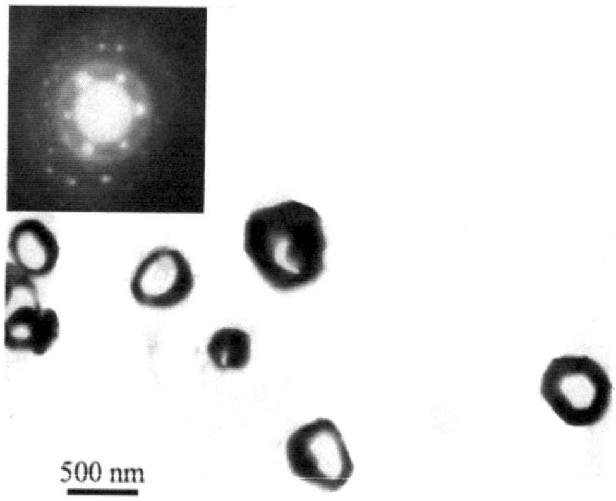

Figure 2 TEM image of the Au nanorings prepared from $HAuCl_4$ (0.2 mmol/L)-SDS (3 mmol/L)-PEG 20000 (5 mmol/L) system after 10 days and inset is electron diffraction pattern of a single Au nanoring lying flat on a TEM grid.

It is documented that the surface plasmon resonance (SPR) band of Au nanostructures strongly depends on the size, shape, and aggregation of the nanostructures (Ah, 2005). Spherical Au particles show an SPR band at approximately 520 nm that is usually red-shifted to longer wavelengths with increasing particle size (Wang and Chen et al., 2005). Figure 3 shows strong SPR band above 800 nm thus providing another piece of evidence for the formation of anisotropic Au particles.

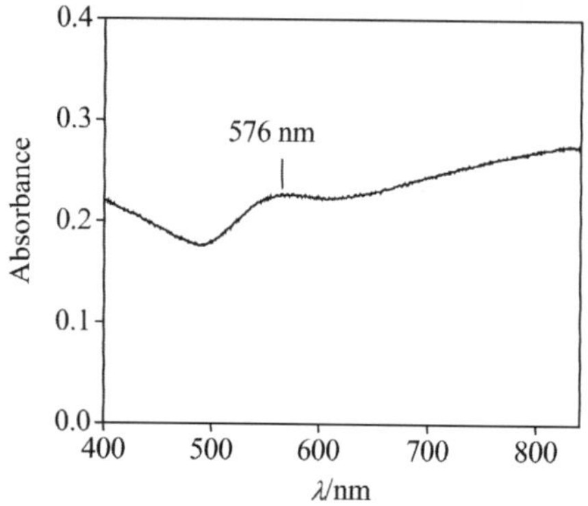

Figure 3 UV-vis spectrum of the Au nanorings prepared from $HAuCl_4$ (0.2 mmol/L)-SDS (3 mmol/L)-PEG 20000 (5 mmol/L) system after 10 days.

In order to investigate the formation of Au nanoring, the evolution of Au nanostructure with different reduction period was trailed by TEM and SEM (Figure 4). The images in the figure 4 show that the reaction period has great influence upon the structures of Au nanoparticles. The Au nanospheres primarily produced in SDS-PEG solution quickly turned into nanoplates and eventually into nanorings. $HAuCl_4$ in reacted solution turned to Au spherical particles (~20 nm) after 6 h (Figure 4 a), and polygonal planar Au nanostructures (~100 nm) such as hexagonal or truncated triangular Au nanoplates appeared after 1 day (Figure 4 b). As the reaction proceeded, the number of small Au nanospheres decreased companying with the increase of hexagon or truncated triangle Au nanoplates after 3 days (Figure 4 c). It was interesting that some hollows emerged in the center of nanoplates after 6 days (Figure 4 d), in fact these mystical hollows loomed up as soon as the third day if someone observed carefully. There occurred lots of ring-like nanostructures after 10 days (Figure 4 e), compared with their nanoplate precursors, the outside vertical angles of these nanorings had become blurred or disappeared.

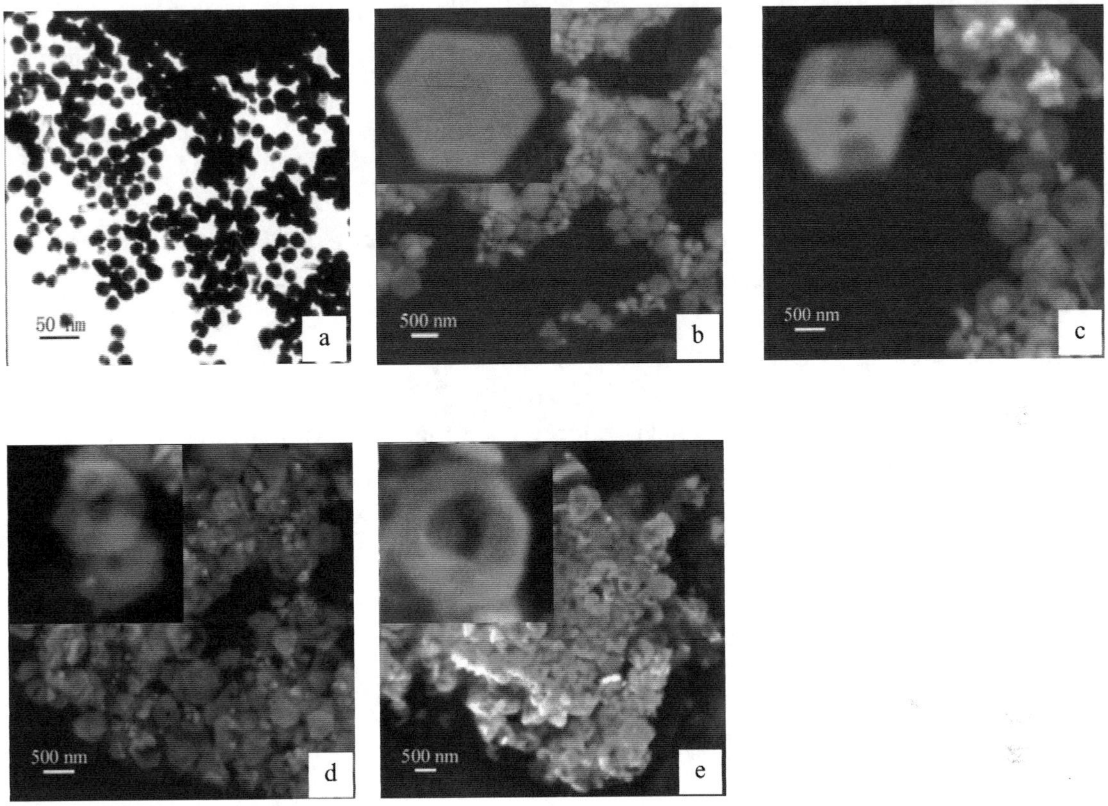

Figure 4 Au nanostructrues obtained from $HAuCl_4$ (0.2 mmol/L)-SDS (3 mmol/L)-PEG 20000 (5 mmol/L) systems. (a) TEM image of Au nanostructrues reduced after 6 h. SEM images of Au nanostructrues reduced after (b) 1, (c) 3, (d) 6 and (e) 10 days.

The system of SDS-PEG has been investigated previously by our research team and other laboratories (Jiang, 2000; Fang, 2004; Fang, 2006). When the concentration of SDS (c_s) was above the first critical concentration (c_1), SDS and PEG could form the "necklace-like" soft clusters acting as soft template to control the morphology of Au nanostructures. In the process, Au particles absorbed by SDS mainly grew along the (111) lattice planes (Kuo, 2004; Kuo and Huang, 2005). Au particles protected by SDS were associated together by PEG chains in solution, which made them aggregate, fuse and self-assemble to form other anisotropic nanostructures (Biggs, 1994). According to Kelven formula, small particles have greater solubility than the big ones in solution, thus the small ones would self-dissolve to disappear and the big ones would grow up. Therefore, the thinner parts of Au nanoplates would self-dissolve first and induce hollows in the center of the nanoplates. According to interface self-contraction principle, all the inside and outside edges of nanorings tend to round and become ring-like structures.

4 Conclusion

In conclusion, soft template composed of SDS-PEG clusters provides a simple and controllable approach to prepare Au nanorings with diameter of 500$\pm$50 nm. In the process, Au nanoparticles grow from Au nanospheres to nanoplates and eventually to nanorings at ambient temperature and atmosphere. These Au nanostructures are homogeneous in size and dispersed well. By controlling the reaction period, we can obtain different Au nanostructures. The prepared Au nanoplates and nanorings could be oriented on the surface of the substrate and might help to explore new applications in the direct electrochemistry of biomolecules such as horseradish peroxidase in the future.

Acknowledgment

We are grateful for the financial support provided by the National Science Foundation of China (20371021).

References

B. NikoobakhtM. A. El-Sayed, Chem. Mater. **15** (2003) 1957-1962.

S. Biggs, P. Mulvaney, C. F. ZukoskiF. Grieser, J. Am. Chem. Soc. **116** (1994) 9150-9157.

C. S. Ah, Y. J. Yun, H. J. Park, W. Kim, D. H. HaW. S. Yun, Chem. Mater. **17** (2005) 5558-5561.

C. H. KuoM. H. Huang, Langmuir **21** (2005) 2012-2016.

C. H. KuoM. H. Huang, Langmuir **21** (2005) 2012-2016.

C. H. Kuo, T. F. Chiang, L. J. ChenM. H. Huang, Langmuir **20** (2004) 7820-7824.

Z. R. Guo, Y. Zhang, Y. Duan Mu, L. Xu, S. L. XieN. Gu, Colloids and Surfaces A: Physicochemical and Engineering Aspects **278** (2006) 33-38.

C. J. Huang, Y. H. Wang, P. H. Chiu, M. C. ShihT. H. Meen, Materials Letters **60** (2006) 1896-1900.

L. H. Pei, K. MoriM. Adachi, Langmuir **20** (2004) 7837-7843.

L. P. Jiang, S. Xu, J. M. Zhu, J. R. Zhang, J. J. ZhuH. Y. Chen, Inorg. Chem. **43** (2004) 5877-5883.

C. C. Li, W. P. Cai, B. Q. Cao, F. Q. Sun, Y. Li, C. X. KanL. D. Zhang, Adv. Funct. Mater. **16** (2006) 83-90.

L. Y. Wang, X. Chen, J. Zhan, Y. C. Chai, C. J. YangL. M. Xu, J. Phys. Chem. B **109** (2005) 3189-3194.

M. C. DanielD. Astruc, Chem. Rev. **104** (2004) 293-346.

M. M. Maye, W. X. Zheng, F. L. Leibowitz, N. K. LyC. J. Zhong, Langmuir **16** (2000) 490-497.

G. S. Metraux, Y. W. C. Cao, R. C. JinC. A. Mirkin, Nano Lett. **3** (2003) 519-522.

N. R. Jana, L. GearheartC. J. Murphy, Adv. Mater. **13** (2001) 1389-1393.

A. V. Simakin, V. V. Voronov, G. A. Shafeev, R. BraynerF. Bozon-Verduraz, Chemical Physics Letters **348** (2001) 182-186.

W. H. JiangS. J. Han, J. Colloid Interface Sci. **229** (2000) 1-5.

X. P. Sun, S. J. DongE. K. Wang, langmuir **21** (2005) 4710-4712.

Y. Cui, Q. Q. Wei, H. K. ParkC. M. Lieber, Science **293** (2001) 1289-1292.

Y. G. SunY. N. Xia, Adv. Mater. **15** (2003) 695-699.

Y. Fang, K. Cai, L. Y. Zong, X. F. LiuM. S. He, Chem. J. of Chin. Univer. **25** (2004) 888-891.

Y. Fang, S. Huang, F. Chen, S. Fu, M. HeG. Ren, Acta Chimica Sinica **64** (2006) 1587-1592.

Technical Experiences and Economic Aspects of Using Microreactors on Production Scale

Dirk Kirschneck, Microinnova Engineering GmbH, Reininghausstrasse 13 a, Graz, 8020, Austria

A lot of research has been done on micro reaction technology. Some companies started to use this technology on production scale level. Microinnova transfers micro reaction technology into production scale. One Microinnova developed liquid-liquid application with a throughput of 3 tons per hour has been started up more than a year ago. Several others are in the engineering stage and will be started up in the next year. Scale-up experiments have already been done successfully for liquid-liquid applications as well as for liquid-gas applications. The examples have proofed technical feasibility. One of the success stories has been published recently. Technical advantages a next step is to evaluate the economic impact.

Results of more than one year production experience with a 3 tons per hour plant will be presented. Achievements and problems will be discussed in detail. The economic impact of two main aims: energy savings and debottlenecking of a production application will be demonstrated. The technical concept and economic impact of a liquid-gas project will be shown. The economic figures are compared with a conventional plant solution.

Finally some general remarks will be given, how to transfer lab scale results into production scale. Micro reaction technology will have an impact on the way of chemical and pharmaceutical production in the future. Some considerations to new strategies will be presented.

Web Page: www.microinnova.com

Consecutive Polymerization Reactions Conducted in Continuous Microfluidic Reactors

Eugenia Kumacheva, Hung Pham, Wei Li, and Stanislav Dubinski. Chemistry, University of Toronto, 80 Saint George street, Toronto, ON M5S 3 H^, Canada

Recently, synthesis of polymer particles in continuous microfluidic reactors allowed excellent control over microbead size, shape and morphology. Precise control of polymerization reactions in time and in a specific location of the microfluidic reactor has not been reported, yet.

We report continuous and scalable microfluidic synthesis of hybrid polymer particles, that is based on microfluidic emulsification of a mixture of monomers and highly controlled consecutive polymerization of these monomers. The first polymerization reaction is initiated by UV-irradiation, the second reaction is triggered by applying heat to the monomer droplets. The resulting method produces polymer particles with narrow excellent size distribution and controlled internal structure.

Measurements Based on Polymerizations in Microfluidic Devices

Kathryn L. Beers, Thomas Q. Chastek, Kazunori Iida, and Derek Patton. Polymers Division, NIST, 100 Bureau Dr., STOP 8542, Gaithersburg, MD 20899

Advanced polymeric materials in applications as diverse as tissue engineering, electronics and personal care products require superior control of a wide range of properties. From molecular properties such as molar mass and chain composition, to the properties of complex mixtures, high throughput and combinatorial techniques are providing researchers access to enormous libraries en route to both improved fundamental understanding of structure-property relationships and better products. The appeal of microreactor technology to this scientific community is three-fold: further reduction of scale in expensive specialty applications; faster, less expensive process integration in laboratory-scale investigations of complex, multi-step manufacturing protocols; and potential discovery of new and improved products from the unique microreaction environment. The technical mission of the NIST Combinatorial Methods Center (NCMC) has two key components: development of new library fabrication techniques and commensurate high throughput and combinatorial measurement methods. Microfluidic device fabrication methods were designed to facilitate rapid redesign and fast in-house production of the chips. Our recent work demonstrates three basic routes to using the micro-environment to prepare both gradient and discrete polymer libraries from radical polymerizations. I will present these three basic approaches as well as measurements that have been integrated into the devices or enabled by the types of library design that are possible.

Microfluidic-Assisted Nitroxide-Mediated Copolymerizations: Influence of the Micromixer Geometry

Carine Rosenfeld, Christophe Serra, Cyril Brochon, and Georges Hadziioannou. Ecole de Chimie Polymères et Matériaux - Laboratoire d'Ingénierie des Polymères pour les Hautes Technologies, ULP - CNRS UMR 7165, 25, Rue Becquerel, Strasbourg Cedex-2, F-67087, France

Microreaction technology is becoming a very important topic of today's polymer reaction engineering. Due to their good thermal and mass transport properties, as compared to conventional labscale batch reactors, microfluidic devices can be used to carry out exothermic and diffusion-controlled reactions such as polymerizations. Recently, we showed the enhancement of the control over the continuous Nitroxide-Mediated Block Copolymerization (NMBC) reactions, carried out in two microtube reactors connected in series with a micromixer in between. In all experiments, the microsystem gives the smallest polydispersity indices in contrast to the batch process. In this work, the influence of the micromixer's geometry on the molecular weight distribution of the copolymers is investigated on different polymer systems. We highlight the fact that the control of the polymerization can be improved just by using very narrow microchannels inside the micromixer, i.e. by decreasing the film thicknesses of the polymer solutions.

Parallel Multiple Droplet Generators

Wei Li[1], Zhihong Nie[1], Piotr Garstecki[2], and Eugenia Kumacheva[1]. (1) Chemistry, University of Toronto, 80 St. George St., Toronto, ON M5S3E6, Canada, (2) Institute of Physical Chemistry, Polish Academy of Sciences, Kasprzaka 44/52, 01-224, Warsaw, Poland

Droplets obtained by microfluidic emulsification find applications in the continuous synthesis of polymer and inorganic colloids. The productivity of microfluidic synthesis depends on the ability to scale up the production of droplets by conducting emulsification in parallel droplet generators. This paper describes the design of a multiple microfluidic droplet generator containing four parallel flow-focusing devices (FFDs), two inlets and a single outlet. Using FFDs with an identical design we scaled up the production of droplets and examined the variation in their size distribution, in comparison with droplets formed in a single droplet generator. We used selective in-situ functionalization of the microchannels to conduct fast throughput screening of the effects of surface energy of the channels on the formation of droplets. Using FFDs with distinct geometries we simultaneously generated droplets with different volumes. This result can be useful in the study of the effect of droplet size on polymerization reactions. The results of this work can be used in parallelization of the production of droplets and in fast throughput screening of the effect of droplet generation, such as surface chemistry in microfluidic emulsification.

Panel Discussion: Commercialization of Micro Process Technology

Geraldine Andrieux, Yole Développement, 45 rue Sainte-Genevieve, Lyon, F-69006, France, Annette A. Johnston, Dept. 07PP NCF3-3, Abbott Laboratories, 1401 N. Sheridan Road, North Chicago, IL 60064-6265, Juergen Brandner, Institute for Micro Process Engineering IMVT, Forschungszentrum Karlsruhe, Hermann-von-Helmholtz-Platz 1, DE-76344 Eggenstein-Leopoldshafen, Germany, Mark M. Disko, Corporate Strategic Research, ExxonMobil Research and Engineering Company, 1545 Route 22 East, Annandale, NJ 08801, Yiton Fu, Corporate Research, Unilever PLC, Colworth House, Sharnbrook, Bedford, United Kingdom, Douglas Galloway, UOP LLC, Des Plaines, IL 60017, Volker Hessel, Professor, Institut für Mikrotechnik Mainz GmbH, Carl-Zeiss-Straße 18-20, 55129, Mainz, Germany, Dirk Kirschneck, Microinnova KEG, Reininghausstrasse 13a, Graz, 8020, Austria, Jan J. Lerou, Velocys, Inc., 7950 Corporate Boulevard, Plain City, OH OH 43064, Frank A. Meschke, ESK Ceramics GmbH & Co. KG, Max-Schaidhauf-Strasse 25, D-87474 Kempten, Germany, Dominique M. Roberge, Exclusive Synthesis, Process R&D, Lonza Ltd., Lonzastr., Visp, Switzerland, Alberto Simoncelli, F&HC Engineering Process Breakthrough, P&G Technical Centres Limited, Whitley Road, Longbenton, Newcastle upon Tyne, NE12 9TS, United Kingdom, Olaf Stange, Ehrfeld Mikrotechnik BTS GmbH, Mikroforum Ring 1, 55234 Wendelsheim, Germany, Katsuhiko Tahara, Resereach & Development Center, Arakawa Chemical Industries, Co. Ltd, 1-1-9, Tsurumi, Tsurumi-ku, Osaka, 538-0053, Japan, and Xini Zhang, Chemical & Pharmaceutical Development, Johnson & Johnson Pharmaceutical Research & Development, L.L.C., 1000 Rt. 202, Raritan, NJ 08869.

This session begins with a short introduction on the use of Micro Process Engineering, followed by a panel discussion. Attendees may interact both with industrial users and solutions providers in analysing performance and cost of existing technology. Discuss will focus on how Micro Process Technology is used in production, and what impact Micro Process Engineering will have on the chemical industry worldwide.

Acoustically driven programmable microfluidic chips for biological applications

Matthias F. Schneider[1] **and Achim Wixforth** [1,2]

[1] *University of Augsburg, Experimental Physics I, D-86159 Augsburg, Germany*
[2] *Advalytix AG, Brunnthal, Germany*

Abstract

A novel approach for the design of a versatile chip-based microfluidic system with unique properties and functionality is presented. Similar to microarrays and in contrast to many existing technologies, the fluid handling is performed on the flat surface of a programmable chip. Fluidic tracks and functional blocks such as valves, dispensers, mixers, and sensing elements are chemically defined using standard lithographic techniques. The pumping of the liquid and adressing of the functional elements are based on electrically excited mechanical surface acoustic waves (SAW), propagating along the surface of a chip. Typical, real existing examples for the biological application of such programmable microfluidic chips are a droplet based total analysis system for polymerase chain reaction and a planar version of a blood vessel for mimicking blood flow on a chip.

Introduction

Reducing the dimensions of macroscopic biological or chemical laboratories is advantageous for the following reasons: The small scale allows for the integration of various processes on one chip analogous to integrated microelectronic circuitry. Thus, manual handling, e.g. transferring reagents from one process step to the next, can be reduced. Such an integration is the prerequisite for a fully automated data management system covering all steps of a given process.

Furthermore, the required reagent volumes are reduced thus saving both material costs and process time as many of the time consuming amplification steps for biological substances can be omitted. Finally, the miniaturization results in enhanced precision by providing more homogenous reaction conditions and in shorter reaction times, as less sample volume is present at higher concentrations.

So far, in most of the microfluidic systems liquids are confined and moved in tubes or capillaries. Usually, the application of such systems is restricted to continuous flow processes [1]. Small amounts of liquid cannot be handled separately in tubes, as these need to be completely filled in order for the pumping mechanism to work properly. Most of the work in macroscopic laboratories, however, is carried out as a batch process. A typical example is the mixing of two reagents or dissolving a substance in a liquid. One would measure the required amounts in separate beakers, and then pour the substances in a third beaker while agitating with a stirrer.

Microfluidic devices for sophisticated lab applications usually comprise different components: First, the small amounts of fluids need to be confined to some kind of containers or reactors, holding specific amounts of the liquid. These containers and reactors are then connected via miniaturized tubes or channels, which are operated by small pumps and valves. The whole system additionally needs to be interfaced to the outside world. The smallness of a microfluidic chip (tube diameters are typically of the order of 100 μm or less), ensures that only tiny amounts of reagents are needed for a chemical or biological reaction, on the other hand, however, it also causes complications that are not of relevance in macroscopic fluid handling systems.

As usual for fluidic problems, one has to first regard the Navier-Stokes-Equation, describing the flow in a hydrodynamic system. It is a nonlinear equation in the velocity components, reading

$$\rho\frac{\partial\vec{v}}{\partial t}+\rho(\vec{v}\bullet grad)\vec{v}=-grad(p)+\eta\Delta\vec{v}+\vec{f}\,. \qquad (1)$$

$\vec{v}$ is the velocity field of the flow, η the viscosity and and ρ the mass density of the fluid. p denotes the pressure that a fluid element experiences from its surroundings and $\vec{f}$ is an externally applied body force driving the flow. The term $\rho(\vec{v}\bullet grad)\vec{v}$ describes the inertia of the fluid element and $\eta\Delta\vec{v}$ marks the viscous term. The interaction between the fluid confined to a 'lab-on-a-chip' and the tube walls leads to hydrodynamic features that usually are comprised into a single number characterizing the flow behavior in a fluid, the Reynold's number *Re*, describing the ratio between the inertial and the viscous term:

$$\mathrm{Re} = \frac{\rho v^2}{l}\frac{l^2}{\eta v} = \frac{\rho v l}{\eta} \tag{2}$$

l denotes a typical length scale in the system under consideration, for example the channel diameter. For a microfluidc system, *Re* is usually a small number, indicating the little importance of inertia in the problem. The most prominent consequence thereof and hence the most important difference to a macroscopic fluid volume is probably the lack of turbulent flows in a microfluidic system. The transition between turbulent and laminar flow is usually occurring at a threshold Reynold's number $Re \approx 2000$. Given $l \approx 10$ µm, $v \approx 100$ µm/sec, and the material parameters of water ($\rho=10^3$ kg / m^3, $\eta=10^{-3}$ kg / (msec)), we end up at $Re \approx 0.001$. For a lab-on-a-chip application, this smallness usually causes severe problems. For instance, mixing of two fluids or stirring a liquid to enhance homogeneity or to speed up a chemical reaction, is a very difficult task for a purely laminar flow system. Also, pumping of a low Reynold's fluid is difficult, as the interaction of the fluid with the vessel walls mimicks a high viscosity. We will show, that many of these obsticles can be overcome by employing our acoustic methods.

Experimental Details

Flow in a microfluidic system can be achieved by the interaction of the fluid and surface acoustic waves (SAW), propagating at the surface of a solid chip. The driving force behind this interaction and the resulting acoustically driven flow is an effect called 'acoustic streaming'. It is a consequence of the pressure dependence of the mass density $\rho$of a fluid, leading to a non-vanishing time average of the acoustically induced pressure. Although acoustic streaming is a well-known effect for a long time [2] in macroscopic, classical systems, little attention has been paid on it, so far, in terms of miniaturzation.

Here, we report on SAW streaming in a fluid residing directly on top of a planar chip. SAW have been first described in combination with earthquakes [3]. Meanwhile, reduced to the significantly smaller nano scale, they found their way into much friendlier fields: SAW devices are widely used for RF signal processing and filter applications and became a huge industry in mobile communication. Also SAW are a vital part in basic research on the nanoscale. Here, the optical and electronic properties of semiconductor nanosysyems have been intensively investigated over the last two decades [4]. SAW are especially convenient to excite on piezoelectric substrates. A well defined wavelength and frequency can be excited if a specially formed pair of metal electrodes is deposited on top of the substrate. Such electrodes are usually referred to as interdigitated transducers (IDT). A high frequency signal applied to such an IDT is then converted into a periodic crystal deformation and if fed with the right frequency

$$f = \frac{v_{SAW}}{\lambda} \tag{3}$$

a SAW is launched. Here, v_{SAW} denotes the sound velocity of the respective substrate, and the wavelength λ is given by the lithograficallly defined periodicity of the IDT. Typical wavelengths of technically exploited SAW range from about λ=30μm at $f \approx$ 100 MHz. If a second IDT was placed downstream the substrate surface, a so-called delay-line would be formed. Both transducers, their design, and the substrate properties thus act as a high frequency filter with a predetermined frequency response. They are lightweight, relatively simple and low cost, and can be produced very reproducible, which explains their massive use in high frequency signal processing like mobile telephony.

Most of the energy propagating in a SAW (usually more than 95%) is of mechanical nature. Viscous materials like liquids absorb a lot of this mechanical energy at the surface of the chip. The interaction between a SAW and a liquid on top of the substrate surface induces internal streaming, and, as we will point out below, at large SAW amplitudes this can even lead to a movement of the liquid as a whole [5]. In Fig. 1, we depict the basic interaction between a SAW and a fluid on top of the SAW carrying substrate:

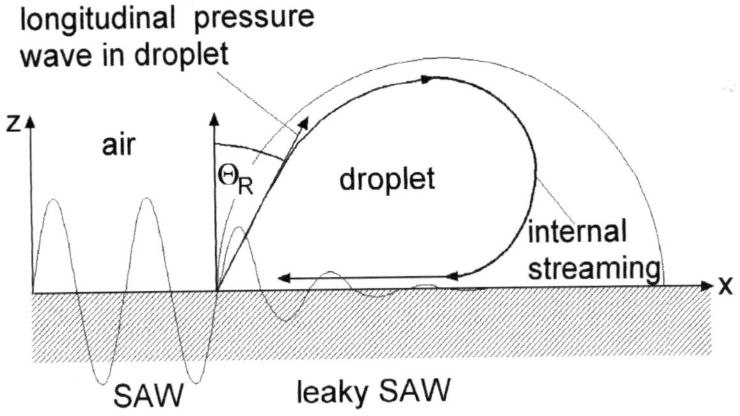

Fig. 1: *Sketch of the acoustic streaming acting on a small droplet on the surface of a piezoelectric substrate. The acoustic energy is radiated into the fluid under an angle Θ_R, leading to internal streaming in the small fluid volume.*

The SAW is approaching from the left and entering the fluid covered region (represented by a droplet in this case) of the chip. There, it becomes attenuated by a viscous damping mechanism, which leads to the excitation of a sound wave in the fluid itself. Phase conservation requires that this sound wave is entering the fluid under an angle Θ_R, very much like for a diffracted beam in optics. SAW streaming then leads to an internal streaming pattern within the droplet, the exact shape of which is determined by Θ_R and the geometry of the fluid volume.

Acoustic mixing

The internal streaming as shown in Fig. 1 can be very efficiently used for mixing smallest amounts of fluid. In Fig. 2, we show a series of two snapshots, about half a second apart, taken for a 50 nl droplet. Here, some fluorescent dye had been deposited at the chip surface just before the water droplet was placed on top. Not only is the dye dissolved by the internal streaming, it is also distributed across the whole volume of the droplet. It should be noted at this point, that the SAW induced streaming still is laminar, as the Reynold's number is so small. The complex flow pattern, however, strongly supports complex material folding lines which in turn facilitate a quick mixing [5]. Moreover, by switching the SAW frequencies and or directions during the mixing, different material folding lines are generated which further improve mixing.

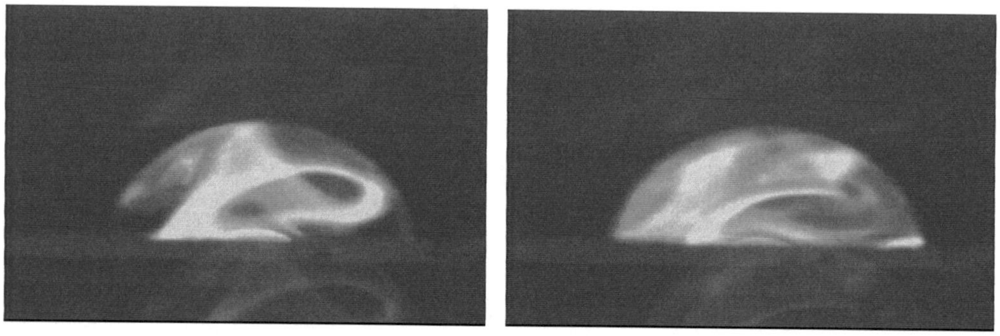

Fig. 2: *SAW induced internal streaming in a small water droplet (side view, approx. 50 nl). A dried fluorescent dye on the surface of the chip is dissolved by SAW agitation, and rapidly fills the whole droplet volume.*

Droplet actuation

For somewhat higher SAW amplitudes, the acoustic streaming effect leads to a strong deformation of the liquid surface and a momentary asymmetry in the wetting angles left and right with respect to the SAW impingement. Especially for small droplets, this leads to a movement and an actuation of the whole droplet into the direction away from the SAW. In this sense, one can regard an IDT on a piezoelectric substrate as an integrated nano pump. Moreover, employing a chemical surface modification, one can define hydrophobic and hydrophillic regions on the chip surface, acting as anchors or fluidic tracks for small droplets. As an example for the versatility of this approch in terms of lab-on-a-chip applications, we show in Fig. 3 a series of snapshots of a programmable microfluidic chip with integrated planar pumps.

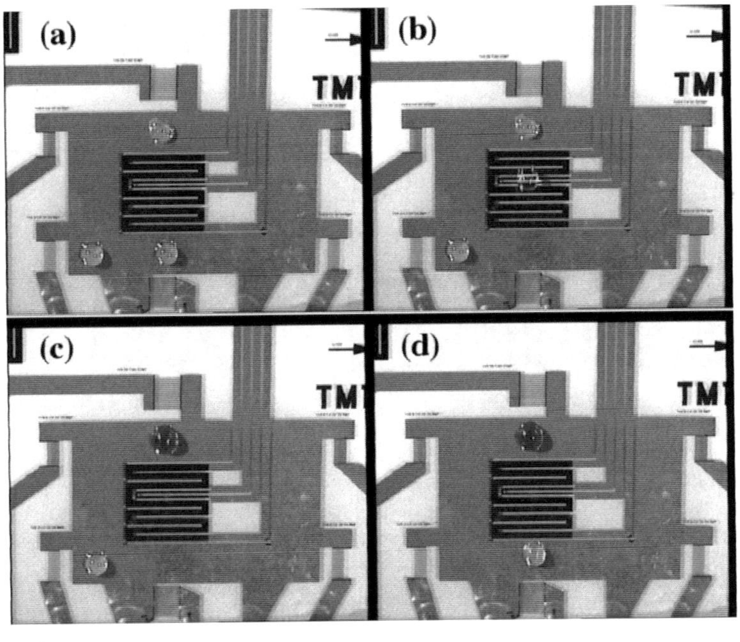

Fig. 3: *A SAW driven microfluidic processor. Three droplets (approx. 100 nl each) are moved ‚remotely controlled' and independently by the nanopumps. (a) through (d) represent a series of subsequently taken snapshots to visualize the movement, and the ‚nanochemical reactions' ocurring when the droplets are merged and mixed by the action of the surface wave. The chip not only contains the nanopumps and the fluidic environment but in the center additional real estate like sensors, and heaters.*

Here, we have used three different droplets of different chemical solutions which exhibit a color change when merged and mixed. It should be noted that while the droplets are actuated and eventually merged, in their bulk, the SAW streaming leads to extremely fast chemical reactions as compared to a diffusion-only driven process.

PCR on a chip

Having a completely programmable microfluidic chip at hand, one can aim towards more complex assays like, e.g., polymerase chain reaction (PCR) on a chip [6]. Here, however, comparably high temperatures are involved during the protocol. Such high temperatures are not compatible with the 'open geometry' of our droplet based fluidics. To avoid evaporation – especially during the 95° Celsius cycle of a PCR process - we cover the aqueous sample solution with a mineral oil layer.

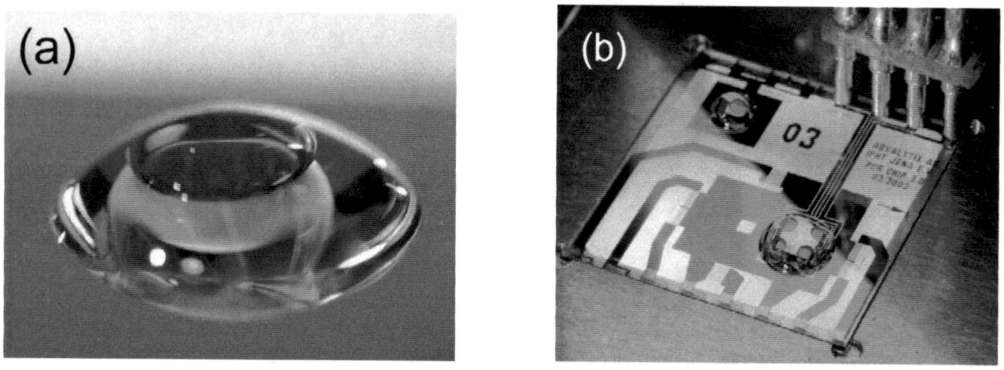

Fig. 4: *(a) virtual liquid test tube' for high temperature application of the planar SAW driven fluidics. To avoid evaporation of the aqueous sample solution (red), the droplet has been covered by a thin mineral oil layer. Both represent a fluid test tube for further processing of the sample. (b) Programmable microfluidic biochip for multi-spot PCR in top view. Apart from SAW driven nanopumps for the fluid actuation, the chip also hosts a heater and a thermometer. The chip is connected to the outside via pogo contacts. Underneath the transparent piezoelectric substrate, a microscope objective is mounted for fluorescence monitoring of the PCR progress employing an intercalating dye.*

The sample contains all the 'PCR-mix' including template, primers, and polymerase for a sucessful amplification of a small amount of genetic material. The sample volume in this case is well below 10 microliters, a 'virtual liquid test tube' as depicted in Fig. 4a. Here, the sample droplet has a red color for better visualization. As the programmable biochip is fabricated employing standard planar lithografical technology, we can also include additional functional elements like thermometers and heaters. The optical transparency and the lack of self fluorescence of the substrate materials used further adds valuable for this lab-on-a-chip application [7]. In Fig. 4b, we show a typical PCR chip. In the lower right corner, an oil droplet is seen, with in this case four different sample droplets, each holding a different sample. From below, an optical system is attached to monitor the progress of amplification using an intercalating dye. For details of the PCR protocol, the sensitivity of the technology, and typical experimental results we refer the reader to [7].

Blood flow on a chip

Studying cell or platelet adhesion (hemodynamics) in small capillaries and arteries is a particularly challenging topic, since it is difficult to be mimicked in an in-vitro system and hard to access by optical or mechanical means. As the diameter of capillaries and arteries become as small as one micrometer, the Reynolds number becomes very low making the pumping mechanism extraordinary difficult. This calls for a new pumping principle when designing a in vitro model to stuy the basic principles taking place during blood flow. Fig. 5 describes the principle of a microflow chamber, which is directly built on a SAW chip (μ-FCC). This system can resembles a type of "artificial blood vessel". The fluid on the chip is confined to a trajectory or a virtual container by a chemical modulation of the surface wettability using hydrophobic/hydrophilic surface functionalization employing soft lithography. The heart of the chip, the nanopump is driven by a surface acoustic wave (SAW) causing the liquid flow by inducing surface acoustic streaming (see Fig. 1). The amplitude of the SAW in the liquid decays exponentially within a characteristic length scale of a few micrometers. Therefore, considering the total length of a typical channel (40mm) the pump basically acts as a point-like source, driving the liquid to flow according to conservation of mass. The flow chamber has no dead volume, allowing for the investigation of even expensive or rare substrates with a volume 100 to 1000 times less than usually required. The particularly small sample volume of only 8μl is probably the most prominent advantage of our nanopump driven planar flow chamber chip. Moreover, the chip components are inert and entirely compatible with biological systems and the system is free of all movable parts making the handling extremely simple. The complete optical transparency of the $LiNbO_3$ substrate allows for individual cells to be tracked over long periods of time, when the μ-FCC is mounted directly onto a fluorescence microscope.

Lanes ranging from a few μm up to 10mm can be realized and the open structure on a free surface allows for direct access to the channel at all times (e.g. for the addition of antibodies or drugs). The technique also provides maximum freedom for mimicking all possible vessel architectures and conditions as they may exist in nature with curved, branched or restricted vessels. Finally, the SAW based pumping system of our chip guarantees for a homogenous, easily controllable laminar flow whose profile covers the complete range of physiological flow and shear conditions from 0 to several $10000s^{-1}$ [8]. Finally, no restrictions are put on the surface functionalization. Artificial lipid membranes, protein coats and even confluent cell layers have been successfully be prepared.

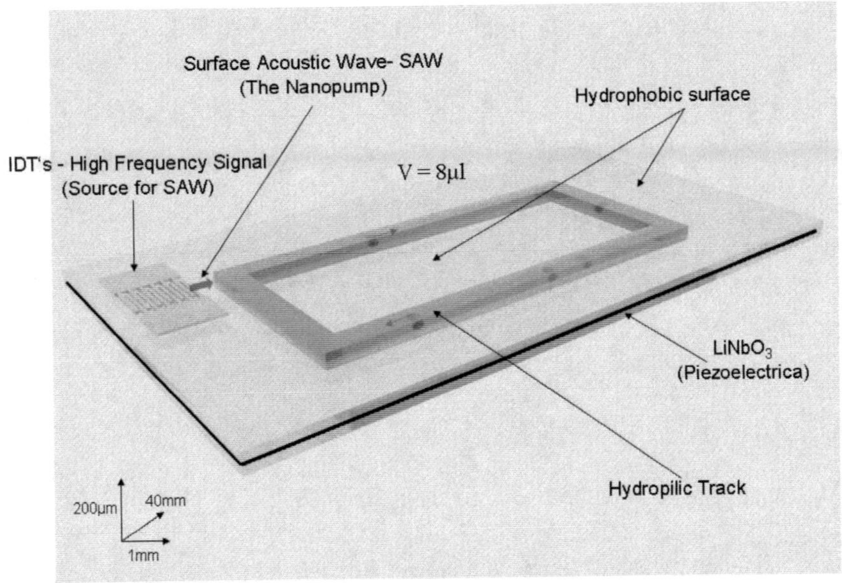

Fig. 5: A SAW driven microfluidic chip. A SAW, excited electrically, eventually interacts with the confined liquid at the solid/liquid interface and drives the liquid to flow (acoustic streaming). The surface wave basically acts as a localized pump, as its mechanical energy is absorbed by the liquid over only a few micrometers. Due to the small scales of this microfluidic system, it creates a homogenous laminar flow (low Reynold's number) along the channel, mimicking the blood flow in arteries or capillaries.

Proteins under Flow

Using this set up, we studied the effect of shear flow on a protein called von Willebrand factor (vWF). It is a glycoprotein being synthesized and stored in endothelial cells, and has found to play an important role for blood coagulation, particularly in regions where the shear rates are high. Under normal conditions, it assembles into multimers (biopolymers) which when stretched can reach sizes as long as 100 µm. The monomeric length is unusually large (~100nm) and contains 2050 amino acids residues. Intuitively, one would expect that blood platelet adhesion is always decreasing with increasing shear force applied to the platelet. Surprisingly, vWF mediated adhesion, on the other hand, is strongly enhanced under high shear-flow conditions [9]. vWF was spread over the hydrophilic track of the flow chamber chip an exposed to various shear flows.

Images of vWF at concentrations typical for human blood ($c \approx 2\ \mu g/ml$) are presented in Fig. 6. At shear rates between $\dot{\gamma} \approx 10$ and 1000 s^{-1}, the biopolymer exhibits a compact conformation (Fig. 6 left). The size of the VWF globules was estimated within our fluorescence setup to be $d \approx 2\ \mu m$, clearly showing that it consists of more than one monomer. This compact conformation remains unchanged as long as the shear rate is maintained below a certain (critical) value $\dot{\gamma}_{crit} \sim 5000\ s^{-1}$. Increasing the shear rate above $\dot{\gamma}_{crit}$ induces a shape transformation of the VWF fibers from a collapsed to a stretched conformation of length $l \approx 15\mu m$ (Fig. 6 right). This transition is reversible, as we observe an immediate relaxation of the protein to its compact conformation when the flow is turned off.

The fact that the transition occurs at such a high shear rate cannot be understood from previous studies on linear chains under good solvent conditions. For example, it has been shown that DNA of roughly the same length as the vWF fibers studied here will exhibit drastic changes of elongation for shear rates as low as 30 s^{-1} at a viscosity of 1cP (i.e. water or the phosphate buffer employed here). Our studies, however, indicate that significant changes in conformation occur only at shear rates $\dot{\gamma} \geq 5000 s^{-1}$. This value is more than two orders of magnitude higher than that reported for DNA. We assume that strong attractive interactions between monomers hold the VWF fiber tightly together, even under strong shear conditions. Therefore, we propose that a single VWF biopolymer forms a compact or collapsed structure similar to a folded protein (Fig. 6 left) [10] .

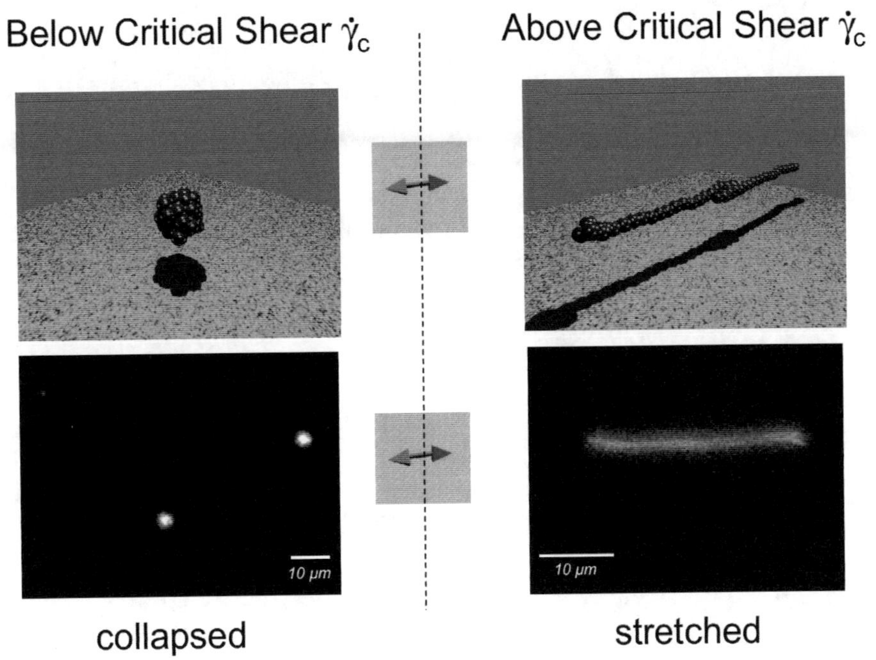

Fig. 6: *Proteins under Flow. The blood clotting protein vWF is exposed to various shear flow conditions. Only when a critical value is exceeded the protein elongates in a reversible fashion. The unusually high shear rates could not be explained by conventional hydrodynamic modells, instead a new theory was developed [11] .*

This explains the counterintuitive observation mentioned at the beginning of this paragraph. Under high shear flow vWF mediates blood platelet adhesion more effetivels. In its collapsed state all binding sites are buried inside the coil. Once the critical shear is exceeded vWF stretches into a long thin fiber exposing all its binding sites. If vWF touches the surface in this conformation, which is highly unlikely in small capillaries, it will immediately be immobilized and serve as a "sticky" grid for free floating blood platelets. Since forces, conformation and function are so closely related, we call this "self organized blood clotting".

Cell-Cell interactions on a Chip

Although our planar flow chamber chip represents an elegant technique for examining many relevant topics, it turned out to perform sub-optimally for cell culture on its surface: Initial attempts to perform cell culture on a planar flow chamber chip were hindered by the fact that the protein rich media influenced the channel architecture due to the protein adsorption on the hydrophobic surface. As a result, this led to an unwanted inhomogeneous flow. To avoid this, we developed a more sophisticated 3d version of our system which easily allows for successful cell culture while retaining all of the advantages of the planar system previously described. We achieved this by a flexible yet stable 3 dimensional architecture by adding "walls" to the channel employing the fully transparent, biocompatible synthetic elastomer Polydimethyl-Siloxane (PDMS).

The usefulness and validity of our system was tested by examining the adhesion of human melanoma cells at various shear rates over time. The results can be seen in Figure 7, which presents our cell culture chip and the PDMS channel. Note that the cells form a confluent layer filling out the entire channel without growing outside of the boundaries. This is evident by the sharp line marking the channel wall. In addition, the cells appear healthy, divide readily and are securely attached to the chip surface. This cultured layer remains completely unaffected by the SAW, which still induces a continuous flow along the channel.

Finally, we were able to demonstrate the use of our system as a cultured micro flow chamber. The individual cells were tracked at three different points in time. Note that the underlying layer of cells is neither moving nor changing in its overall appearance, clearly indicating its integrity. This experimental finding demonstrates the suitability of our chip to be used as a type of vascular model and further shows the ease in which the flow can be observed, attributable to the transparent design.

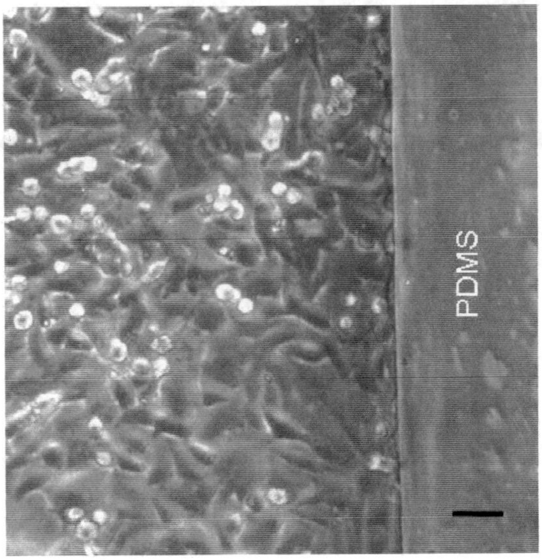

Fig. 7: Cell-Cell Interactions on a Chip. Cells are cultured directly into the microfluidic channel using PDMS walls. The cells form a confluent layer over the entire channel and do not grow across the boundaries (PDMS walls). Cells flowing over the intact and immobilized confluent cell (brighter appearing cells). Thanks to the optical transparency of the piezoelectric substrate and the continuous channel design, the cells can be tracked for an unlimited amount of time. The scale bar represents 50μm.

Summary

In summary, we described an unconventional technique to manipulate smallest amounts of liquid on a chip. Employing surface acoustic waves on a piezoelectric substrate, we are able to actuate individual droplets along predetermined trajectories, or induce acoustically driven internal streaming in the fluid. This internal acoustic streaming can efficiently be used to agitate, mix and stir very small liquid volumes, where the low Reynold's number usually only allows for diffusive mixing. As for typical applications, we described a programmable microfluidic chip for droplet based assays, to perform high resolution microliter polymerase chain reaction (PCR). The technique is equally well suited to actuate or agitate small amounts of liquids either in closed volumes or in an open, droplet based geometry. An example for such a closed though 'open' volume is given in the context of a model for the investigation of blood flow on a chip. Here, we showed the feasibility to even grow a cell culture on predertemined areas of the chip, representing a flat version of a blood vessel. The shear force induced mechanical activation of an important biopolymer for wound healing and cell adhesion is one of our latest unpreceded applications of our quite universal SAW fluidic chip. The combination of the SAW actuated droplet based fluid handling and SAW driven fluidics in closed volumes opens a wide field of many different applications. Many more applications, and many more visualizations of the technology described above can be looked up on either http://www.Advalytix.de or http://www.physik.uni-augsburg.de/exp1.

This work would have been impossible without the combined efforts of many people involved, both at the University of Augsburg, as well as at Advalytix AG. Some of them are listed in the references below, others needed to be left unaccounted for in this article for the sake of limited space. Their valuable contributions, however, are gratefully acknowledged, in any case. Financial support was provided by the german minister for research and technology (BMBF), the Bavarian Science Foundation, and the German Science foundation DFG under SPP 1164, and SFB 486. M.S. likes to thank the Fond der chemischen Industrie for financial support.

References

[1] A recent very comprehensive review of lab-on-a-chip devices and technologies can be fouind at: O. Geschke, H. Klank, P. Telleman, eds. : *Microsystem Engineering of Lab-on-a-chip Devices*, Wiley-VCH, Weinheim (2004)

[2] Eckart, C., Phys. Rev. **73**, 68 (1948),

see also Nyborg, W. L., in W. P. Mason, *Physical Acoustics 2B: Properties of polymers and nonlinear acoustics*, pp. 265-331, New York:Academic Press (1965).

[3] Lord Rayleigh, Phil. Mag. **10**, 364–374 (1905).

[4] Achim Wixforth , *"Interaction of surface acoustic waves, electrons, and light"* in *"MICROWAVE ACOUSTIC WAVE DEVICES FOR WIRELESS COMMUNICATIONS AND SENSING",* Int. Journ. of High Speed Electr. and Syst., Vol. **10**, No. 4 (2000), 1193-1227 (World Scientific Publishing Company)

[5] A. Wixforth, C. Strobl, Ch. Gauer, A. Toegl, J. Scriba, and Z. v. Guttenberg Anal. Bioanal. Chem **379**: 982-991(2004)

[6] A. Toegl, R. Kirchner, C. Gauer, and A. Wixforth Journal of Biomolecular Techniques, **14**:197–204 (2003)

[7] Z. Guttenberg, H. Müller, H. Habermüller, A. Geisbauer, J. Pipper, J. Felbel, M. Kielpinski, J. Scriba, and A. Wixforth, Lab Chip **5**, 308 (2005)

[8] G. A. Trusky, F. Yuan, and D. F., in *Transport Phemomena in Biological Systems* (Pearson Education, Inc., New Jersey, 2004).

[9] Z. M. Ruggeri, J Thromb Haemost. **1,** 1335-42 (2003).

[10] S. W. Schneider, S. Nuschele, A. A. Katz, A. Wixforth, and M. F. Schneider, *PNAS* (2006) **under Revision.**

[11] A. A. Katz, M. F. Schneider, S. W. Schneider, A. Wixforth, and R. R. Netz, Phys. Rev. Lett. **97**, 138101 (2006)

Generation of nanoliter fluid segment patterns by different methods of segment merging

J. Michael Köhler [1)], P. Mike Günther, Mark Kielpinski [2)], Karin Martin[3)], Martin Roth[3)], Josef Metze [4)] and Thomas Henkel [2)]

1)Technical University of Ilmenau, Institute for Micro and Nanotechnologies/Institute for Physics

2) Institute of Photonic Technologies, Jena

3) Hans-Knöll Institute Jena

4) Institute for Bioprocess and Analytical Techniques iba Heiligenstadt

correspondence:
Prof. Dr. Michael Köhler, Techn. Univ. Ilmenau, Institute f. Micro & Nanotechnologies, PF 10 05 65, D- 98684 Ilmenau, Germany, Tel. 49- (0) - 3677 - 69 - 3655, Fax - 69 - 3179 , michael.koehler@tu-ilmenau.de

Abstract

Regular pattern and complex sequences of fluid segments can be generated in micro fluidic systems. Either the fluid patterns are formed directly using a combination of injectors and junction in a micro fluidic network or they result from the merging of preformed series of fluid segments. In both cases, the resulting pattern is not only a simple superposition of original sequences but is determined additionally by the coalescence behaviour of the segment liquids. So, the resulting sequences contain fused segments with higher volume and compositions depending on the individual original segments to be fused. In general, the formation of regular patterns is better reproducible in case of merging of preformed simple sequences. Segment sequences are characterized by micro flow-through photometric measurements and can be distinguished by the size, distance and order of segments.

Key words

fluid segments, microfluidic network, sequence merging, two phase flow, segment pattern

Introduction

The definition of well defined small reaction volumes is one of the most important tasks of micro reaction technology. The spontaneous formation of droplets of one liquid during injection of this liquid into an other liquid immiscible with first can be used for this purpose. Droplets of nanoliter, picoliter or even smaller volumes are generated and produced for long time in emulsion technologies. Such, liquid two phase systems are also generated, manipulated and applied in microchannels for several micro reaction application [1-3]. The application of multi-phase flow is preferred in miniaturized chemical flow-through processes due to the possibility of realizing narrow residence time distributions. In contrast to classical emulsions, the formation and transport conditions of droplets inside micro tubes and channels offer the possibility of distinguishing the individual droplets. This becomes possible by the linear order of droplets. Beside the order, also the distance of droplets remain constant, if their volume is large enough for filling the whole cross section of a tube or micro channel. So, a regular series of liquid droplets is obtained which are embedded in a stream of the immiscible carrier liquid [4]. This sequence of droplets is called a micro segmented flow, the droplets are fluid segments and are well characterized by their position inside the one-dimensional liquid arrangement.

The application of nanoliter fluid segments is of interest for chemical synthesis [5,6], analytical and kinetic studies [7,8] as well as for miniaturized screening experiments [9]. The ideal plug flow transport of fluid segments in micro tubes and micro channels allow to identify and to address the individual segments. So, it can be used for automated analytical procedures as well as for miniaturized combinatorial experiments. The definition of different concentration patterns is necessary in all mentioned applications. This can be achieved either by mixing of reactant solutions before segmentation, by two or multistep injection technique or by merging of droplets. Here, we report about strategies for generation of concentration patterns by merging of segment sequences and by segment fusion.

Experimental

The generation of complex sequence pattern of nanoliter fluid segments is possible by combination of injectors and T junctions in a microfluidic network. The formation of such segment patterns was studied in modular arrangements of PTFE (polytetrafluorethylene) tubes in combination with PTFE T-junctions and in chip devices for injection and segment fusion (Fig. 1a [4]). The formation of pattern with certain concentration distribution were studied by combination of glass/glass chip injectors with static micro mixers [10] in a modular microfluidic arrangement (fig 1b) and in an integrated chip device containing Y-junction mixers with injectors [11] and mixing loops (fig. 1c).

The obtained segment sequences were monitored by micro flow-through photometry allowing a resolution of size and distance in segment measurements better than 1 % up to segment frequencies of more than 10 Hz. Therefore, tube photometers including a LED, an aperture element for illumination and a detector diode are applied. This photometers are working with a relative precision of photometric readout of about 2% down to integration times of 0.2 ms. The formed sequence pattern can be well characterized by the distribution of concentration (absorbance), segment size and distance to the next segment. On-line micro spectrophotometry was used for characterization of changing concentrations of dye mixtures (Fig. 2). The segment parameters can be extracted from the original absorbance-time data sets. Complex patterns are reflected by the formation of different subpopulations in size-absorbance and size-distance plots. The quality of the formed sequence patterns is marked by the parameter scattering inside the single subpopulations of segments.

Results and discussions

Segments in sequences can be distinguished by size, distance and content (fig. 3).

There are several different strategies (fig. 4) for the generation of segment sequences with certain patterns:

1. Shifting of composition of liquid before segmentation (one injector)

2. Shifting of composition of a liquid injected into preformed segments (two injectors)

3. Shifting of flow rate ratio of a first liquid stream generating the segments and a second liquid stream injected into the segments (two injectors)

4. Merging of segment sequences without coalescence segments

5. Merging of segment sequences with coalescence of segments

Method 1 and 2 can be realized by two input channels conducting liquids to a common injector or by a micro mixer inside the input channel. Method (1), (2) and (3) generate segment sequences with a stepwise variation of concentration inside segments. Slow sweep rates of initial concentrations or flow rates result in small steps in composition between the single segments. Method (3) causes a variation in segment size beside the variation in segment composition, if the flow rate of the injected liquid at the second injector is lower than the flow rate of injected liquid at the first injector (3a). An alternating sequence of segments of the liquid injected at the second injector and segments of mixed composition results if the flow rate at the first injector is lower than the flow rate at the second (3b). In method (2) as well as method (3) two injectors are applied in a linear (serial) arrangement.

The injection of a second liquid in preformed segments has the advantage of avoiding any problems of fluidic dispersion, if concentrations are changed by shifting flow rates. But, problems of reproducibility can arise from fluctuations in the conditions of droplet release. Each shift in the time interval between droplet release at the first injector causes shifts in segment distance and therefore different sizes of injected volume at the second injector.

Method (4) and (5) use at least two injectors in a parallel channel configuration. A superposition-like pattern is created if no coalescence of segments take place. A regular 1:1 alternating pattern is formed if both input sequences have the same sample frequency and are applied at same flow rate (4a). Sequences in form of n:1 or 1:n pattern are formed in case of different segment frequencies in the input sequences or in case of different flow rates during sequence merging (4b). Sequence pattern with alternating n:1 and (n+1):1 motives are obtained if the input flow rates or input segment frequencies are applied in ratios with deviation from the simple n:1 ratio (4c).

Much more complex segment patterns are formed, if coalescence takes place. The resulting segments are different in size and composition. The formed sequences can contain segments of the original size and composition as well as larger segments with a compositions resulting from the mixing of the original segments (5). The distributions of segments sizes and segment compositions depend on the number of merging steps, on the topology of fluid network and on the coalescence tendency of the liquids. Much more complexity arises in case of introduction of splitting operations in the fluid network or in case of fluctuations in size, distance and composition of original fluid segments.

The simplest method of characterization of segment populations is are length/period plots. Therefore, the duration of passage of segment through the optical measurement channel of the flow-through photometer and the time between the end of one and the beginning of the following segment are registered. Homogenous populations are marked by a single cloud of points in the plots (fig. 5a). It can be denser or less dense (in case of higher fluctuations in segment release, fig. 5b). More complex pattern are obtained after merging or splitting procedures with segment sequences in fluid networks (example in fig. 5c). The character, the number and distribution of subpopulations is mainly determined by the topology of the fluidic network. The scattering of points is caused mainly by the interaction of different fusions or splitting events. The generation or and transport of segments in one part of a fluid network is always influenced by the arise of pressure pulses by segment generation or fusion in other parts. Particular sensitivity is developed in systems with ring structures (branching and reunification of channels).

The merging of preformed sequences allows the generation of certain segment pattern with higher regularity. The segments sizes and distances of the original sequences can be defined independent on each other. The photometric measurements reflects high regularity of resulting pattern if the two original sequences are fed synchronously in a sequence merging. An alternation of two segments with higher concentration and one segment with lower concentration of a dye is shown in fig. 6a. Peaks of united segments of lower and higher concentration are obtained, if the phase shift between the synchronized input sequences is lowered below a critical threshold value (fig. 6b). Transitions between pattern with separate segments and united segments are observed in case of small frequency differences between the input sequences (fig. 6c). The flow-through photometry allows the automatic registration of larger sequences and supplies size, distance and absorbance values for each segment. The effect of pattern formation by sequence merging is also be demonstrated by patterns formed from original segments of two different dye solution. Alternating original segments of both dyes (fig. 7a), alternation between original segments of one type and united segments (fig. 7b) and more complex patterns of original and united segments (fig. 7c) are obtained in dependence of the flow rates during sequence merging.

The unification of segments in simple T- or Y junctions is strongly affected by fluctuations inside the both input channels. A higher regularity can be achieved by special devices for pair-wise fusion. Such a channel structure was implemented in glass/glass chip technology. An additional channel with two nozzle structures connects the both inlet channels and ensure the release of a segment in the fusion area only after unification of two segments each coming from one inlet channel (fig.

8). The regulation occurs by coupling the pressure in both inlet channels.

Micro fluid segment patterns are of particular interest for screening and test series with step-wise varied concentration. Therefore, the generation of well-defined concentration pattern is a urgent demand for the application of segment sequences. Segment series with tiny differences in dye concentration can be generated by use of double injector devices. Fig. 9 gives an example of an on-line micro spectralphotometric measurement during the generation of segments with different ratios of two dyes. The microfluidic conditions affect strongly the quality of concentration patterns. In case of high fluctuations, a large scattering in concentrations and sometimes different droplet populations are observed (fig. 10a). Such series can not be used for parallelized or serial chemical or biological experiments. Regular concentration variations can be obtained in case of well-controlled fluidics (fig. 10b). Triangular concentration pattern with large periods can be generated by modular arrangements as well as by integrated devices. Modular systems with certain delay between mixing and segment formation can not be applied for generation of fast changes in concentrations for high frequency concentration patterns. But, this type of patterns can be convenient formed by use of integrated devices (fig. 10c).

Conclusions

There exist different strategies for the generation of fluid segment patterns. The application of simple T-junctions leads to a higher sensitivity in relation to fluctuations. Certain statistical distribution in the fusion events result in more distributed sizes and compositions of segments in finally formed sequences. Chip devices with an internal pressure regulation support the formation of a strong controlled pair wise fusion, if the flow rates are well adapted. In alternative, regular pattern can easily be prepared by merging of preformed sequences. The resulting pattern is determined by the properties of the original sequences, by the flow rate ratio during sequence merging and by the interface properties of the applied liquids. In result it could be shown, that nanoliter fluid segment pattern of very different character can be produced in dependence on flow rate ratios of carrier liquid and embedded liquids, on coalescence behaviour of originally formed segments and on the topology of the used fluidic network.

Acknowledgement

We thank J. Albert and M.Sossna (IPHT Jena), F. Möller and S. Schneider (Ilmenau) for technical assistance. Valuable discussion and support by G.A. Gross (Ilmenau), K. Lemke and A. Grodrian (Heiligenstadt) and financial support by the BMBF (project Serizell) are gratefully acknowledged.

References

[1] I. Kobayashi, M. Nakajima: Generation an Multiphase flow of emulsions in microchannels, in: N. Kockmann (ed.): Micro Process Engineering (Weinheim 2006), 149

[2] J. R. Burns and C. Ramshaw, Lab on a Chip 1 (2001), 10

[3] J.M. Köhler, Th. Henkel, A. Grodrian, Th. Kirner, M. Roth, K. Martin, J.Metze, Chem. Engn. J. 101 (2004), 201

[4] T. Henkel, T. Bermig, M. Kielpinski, A. Grodrian, J. Metze, J.M. Köhler: Chem. Engn. 101 (2004), 439

[5] A. Günther, K.F. Jensen, LabChip 6 (2006), 1487

[6] J.M. Köhler, Th. Henkel, A. Grodrian, Th. Kirner, M. Roth, K. Martin, J. Metze, Chem. Engin. J. 101 (2004), 201

[7] G. Christobal, L. Arbouet, F. Sarazzin, D. Talaga, J.-L. Bruneel, M. Joanicot, L. Servant, Lab Chip 6 (2006), 1140

[8] B. Zheng, J.D. Tice, R.F. Ismagilov, Anal. Chem. 76 (2004), 4977

[9] K. Martin, Th, Henkel, V. Baier, A. Grodrian, Th. Schön, M. Roth, J.M. Köhler, J. Metze, Lab Chip 3 (2003), 202

[10] T. Kirner, J. Albert, M. Günther, G. Mayer, K. Reinhäckel, J.M. Köhler: Chem. Engn. J. 101 (2004), 65

[11] I. Shestopalov, J.D. Tice, R.F. Ismagilov, Lab Chip, 4 (2004), 316

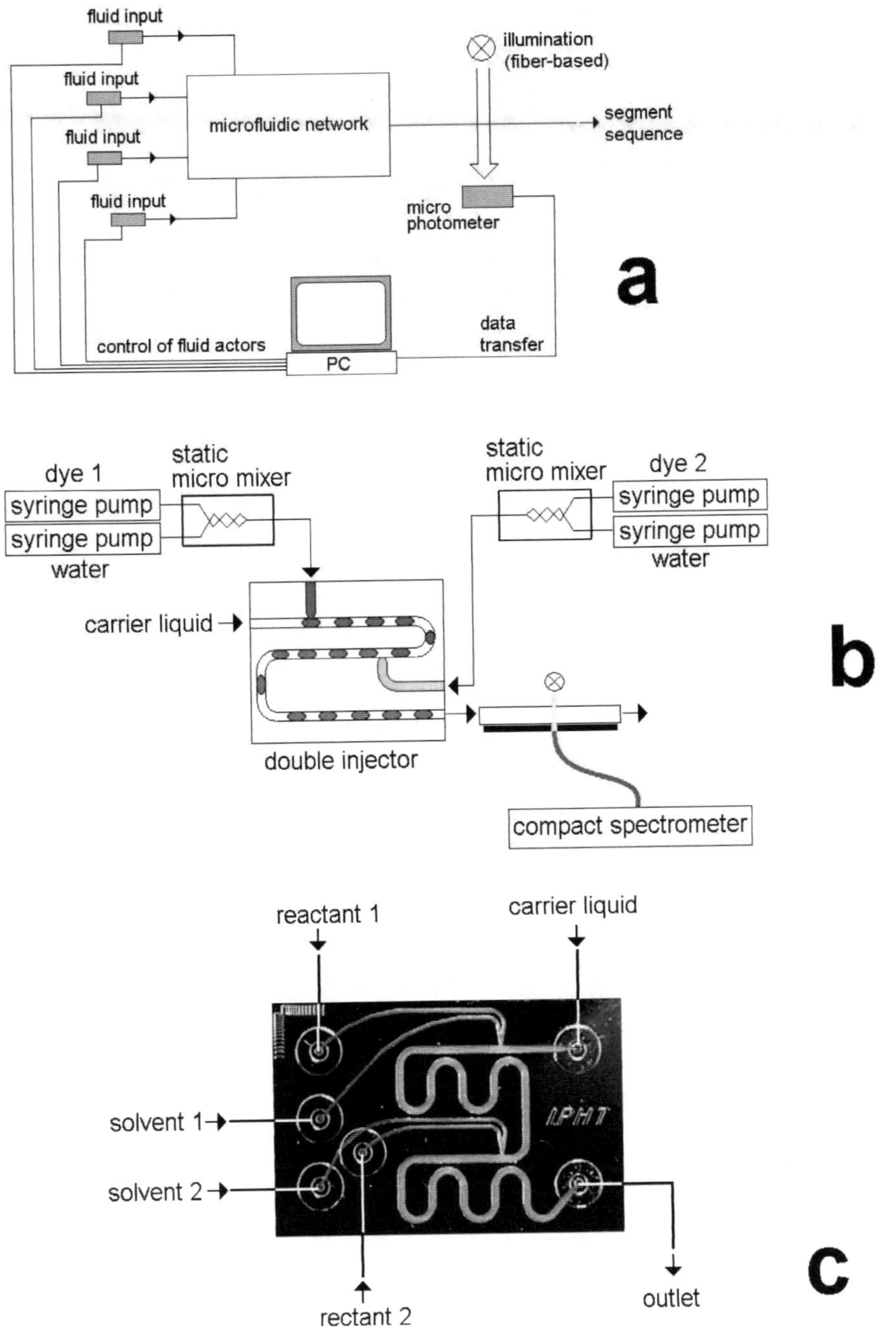

Fig. 1

Fluidic arrangements for generating of segment patterns: a) scheme of PTFE T-type injectors and tubing, b) modular arrangement with static mixers and double injector, c) integrated mixing and injecting chip (glass/glass), d) detail

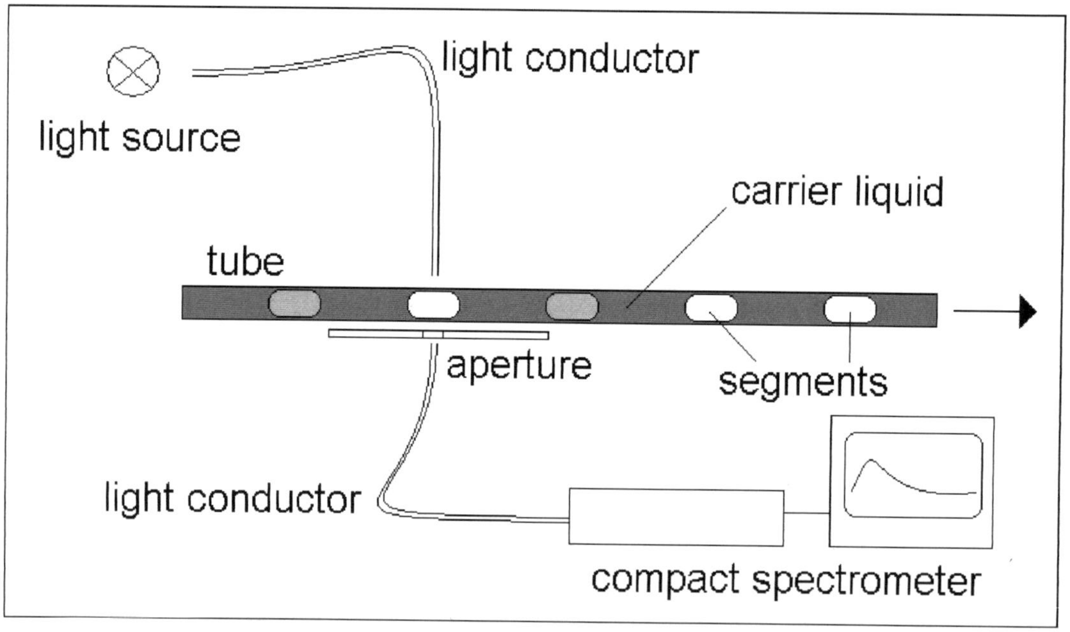

Fig. 2

Fiber optic arrangement for spectral characterization of segment liquids

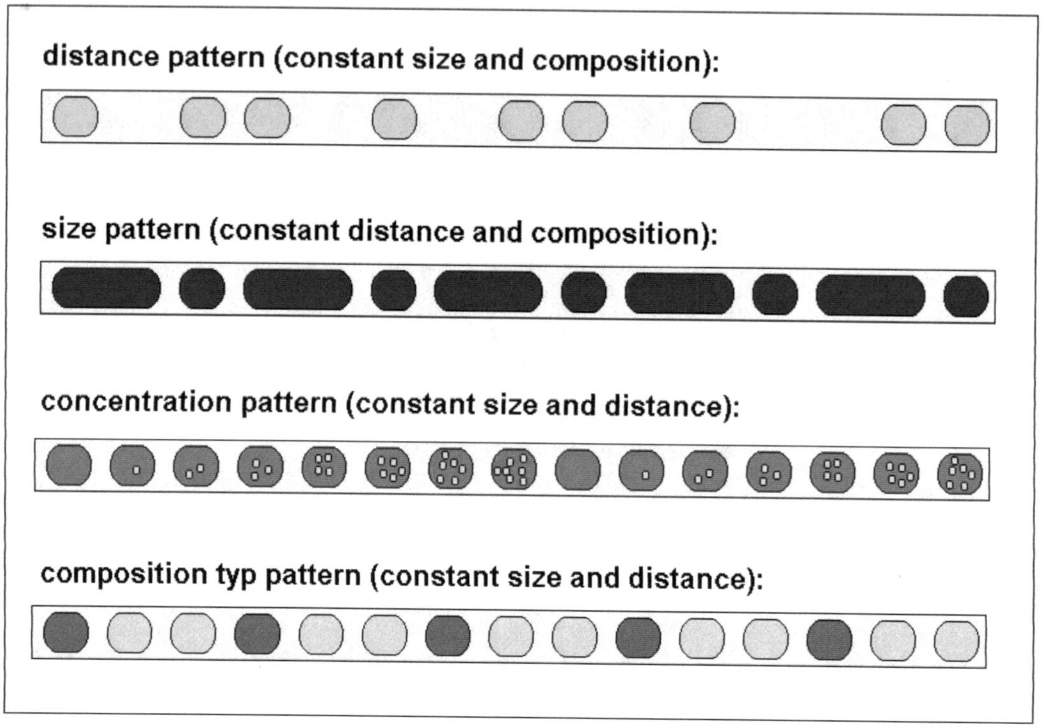

Fig. 3

Categories of features for segment sequence patterns

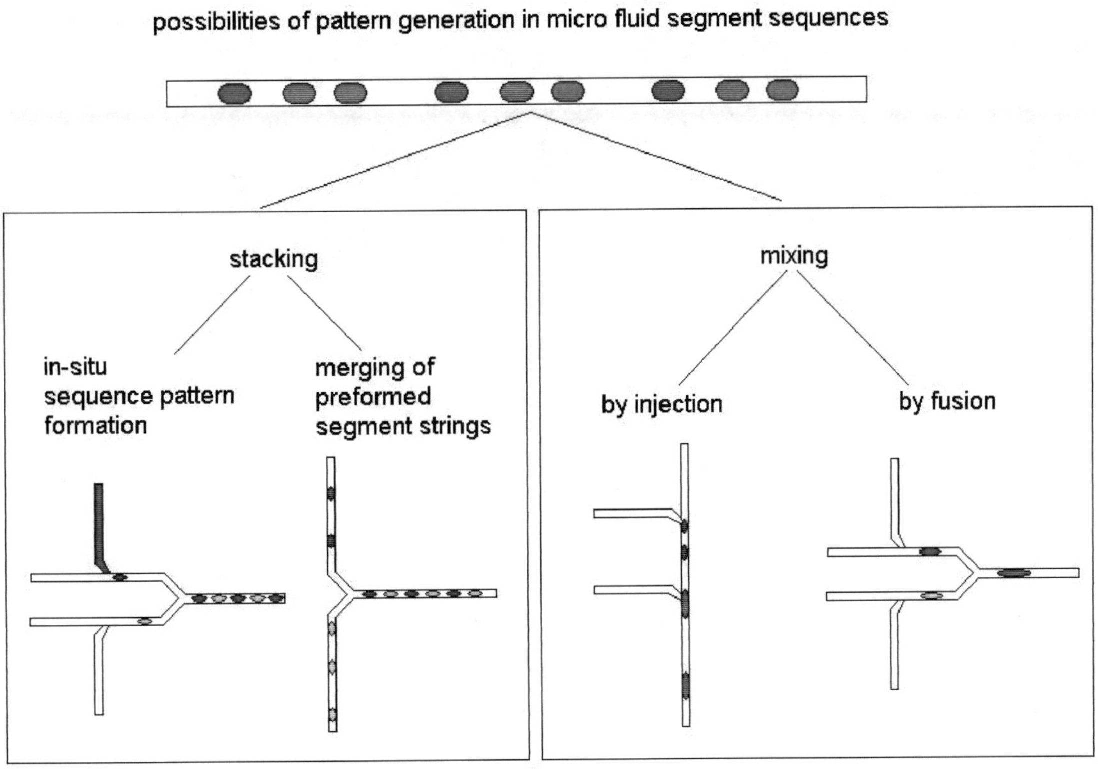

Fig. 4

Operations for generation of sequence patterns of micro fluid segments

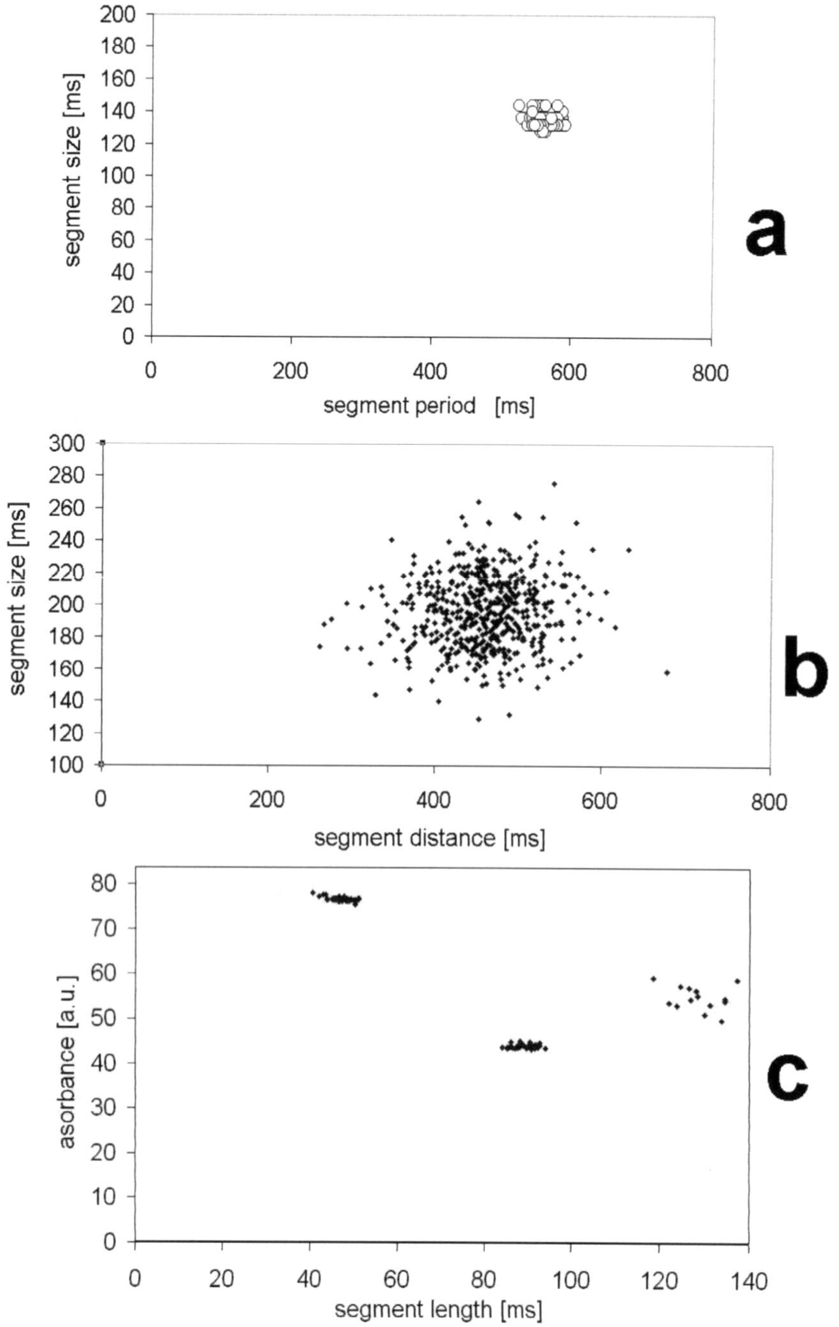

Fig. 5

Characterization of segment populations by segment size/segment period plots (characterized by time of transition at the position of flow-through photometer: a) uniform population (low scattering of values), b) population with high scattering of sizes and distances, c) subdivision in subpopulations (in-situ merging of two sequences in an arrangement with 3 T-junctions)

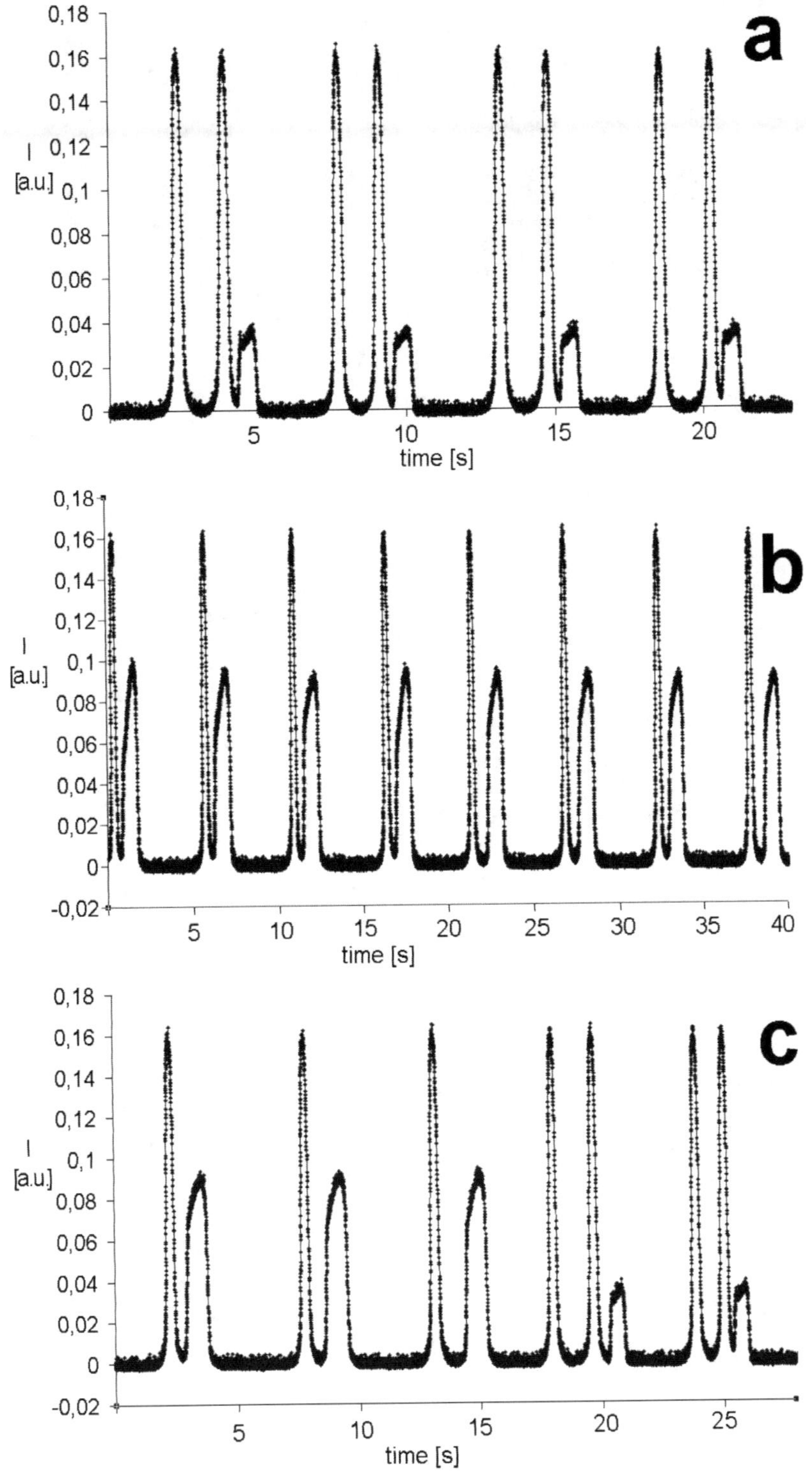

Fig. 6
Photometric signal of segment pattern after merging of two preformed sequences of the same dye in two different concentrations: a) without coalescence, b) with partial coalescence, c) transition between coalescence and non-coalescence due to a small frequency ratio shift

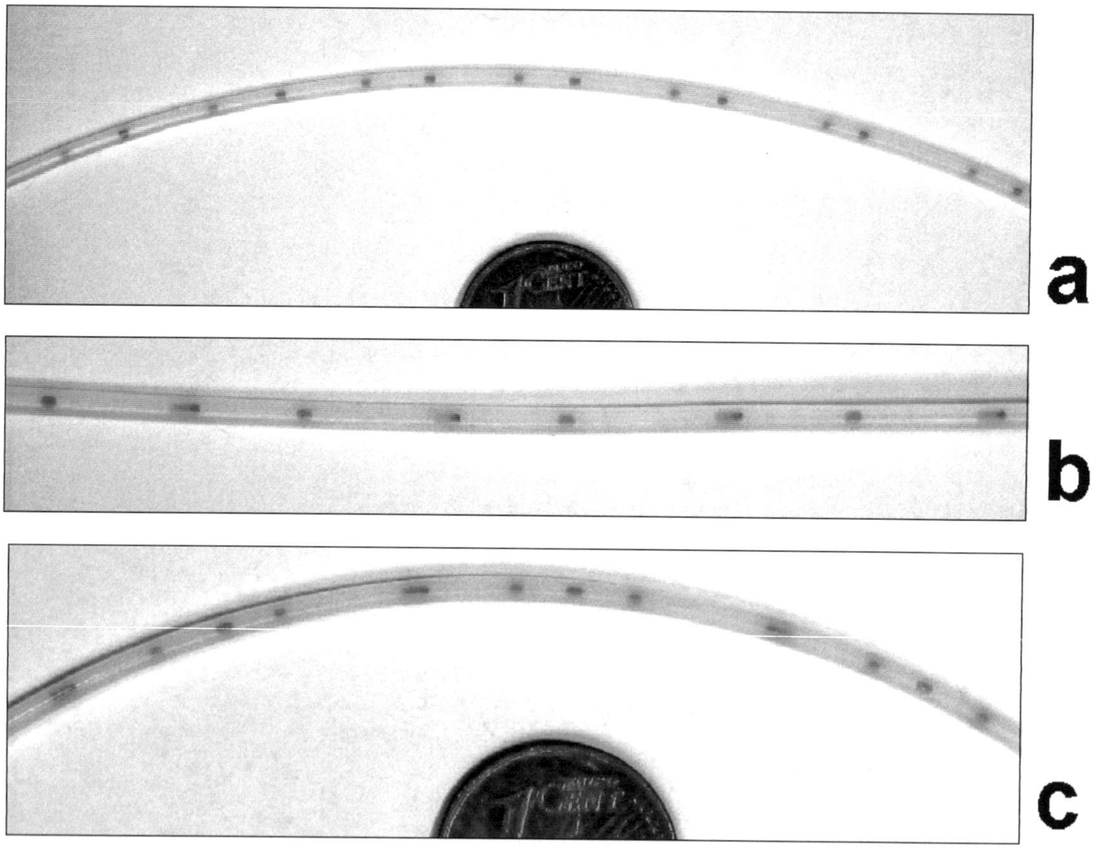

Fig. 7
Sequences obtained from two segment series of solutions of two different dyes: a) 1:1 pattern, without coalescence, b) partial coalescence (alternating original and fused segments), c) complex pattern (three smaller original segments alternating with two larger segments after fusion)

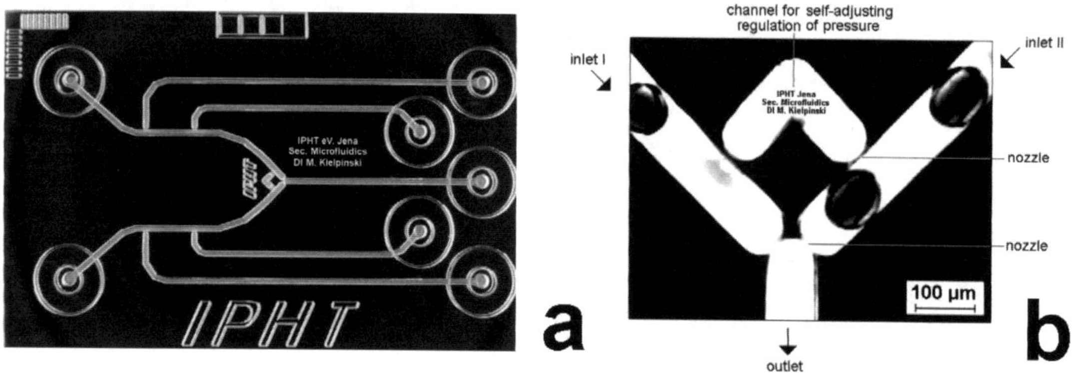

Fig. 8
Micro device for self-adjusted pair-wise fusion: a) chip device and arrangement, b) detail of fusionator chip

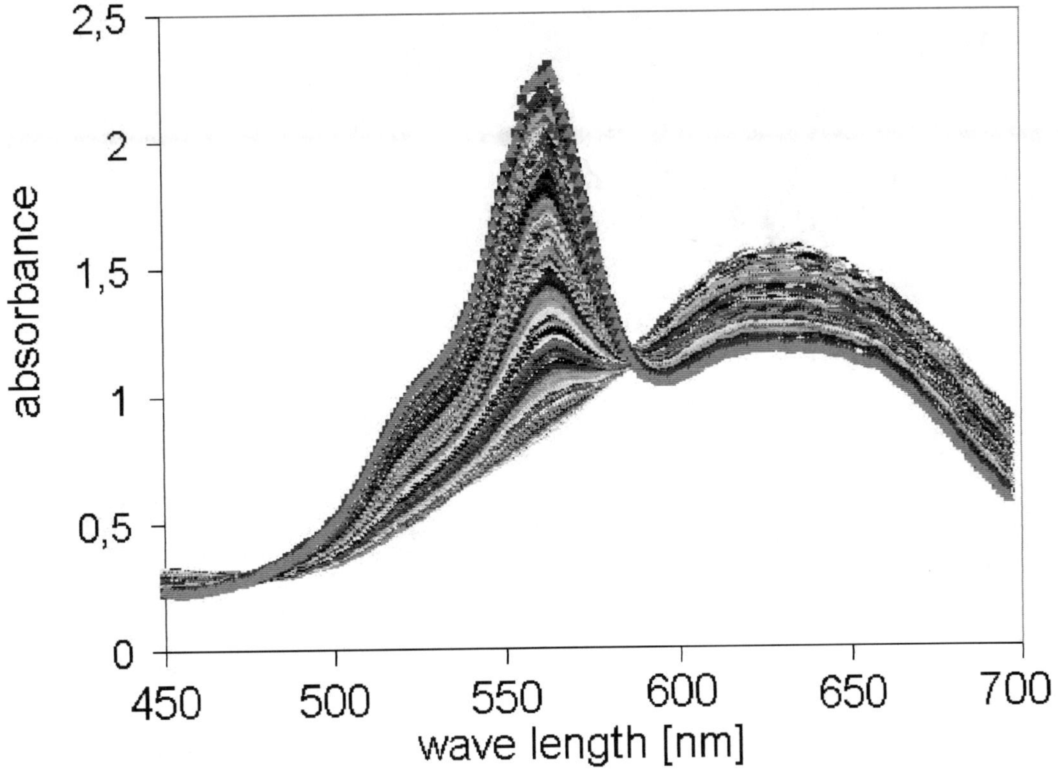

Fig. 9
Formation of a segment sequence with step-wise varied concentration ratio of two dyes: optical absorption spectra obtained by in-situ micro flow-through spectral photometry

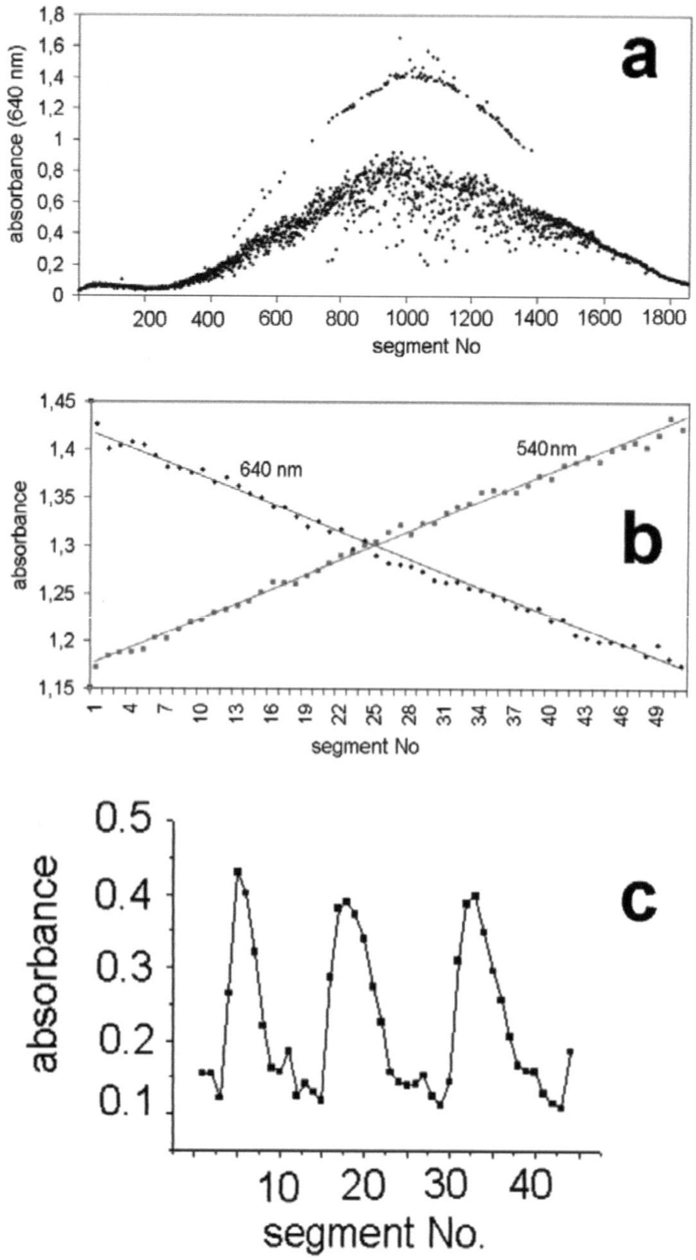

Fig. 10
Optical absorption of segments at constant wave length during experiments for generation of sequences with step-wise varied concentration (each symbol represents one segment) : a) large scattering and formation of subpopulations in case of high fluctuations b) segments in a population with small concentration steps obtained in case of well-controlled flow rates and injection conditions (modular arrangement with two static micro mixers and double injector , c) sequence with 3 concentration maxima and three concentration minima and larger concentration steps between segments obtained by experiments with integrated mixer/injector chip

Lambda DNA Thermal Migration in a Microchannel

Jennifer Kreft, Institute of Physics, Academia Sinica, 128 Academia Rd Sec. 2, Nangang, Taipei, Taiwan and Yeng-Long Chen, Institute of Physics and Research Center for Applied Science, Academia Sinica, 128 Academia Rd Sec. 2, Nangang, Taipei, Taiwan.

The Ludwig-Soret effect, the migration of a species due to a temperature gradient, has been extensively studied without a complete picture of its cause. Many particle and polymer solutions have been used to study the phenomena; here we investigate the dynamics of DNA confined to a microchannel and subjected to a thermal gradient using a combination of Brownian dyanmics and the lattice Boltzmann method. We observe that the DNA molecules will migrate to colder regions of the channel, an observation also made in the experiments of Duhr, et al. In fact, the thermal diffusion coefficient found for a range of temperature gradients agrees nearly quantitatively with the experimental value. Furthermore, we use the simulation to understand how the interaction of the DNA with the solvent leads to thermal migration via the thermal fluctuations-fluid momentum flux coupling.

SPECIFIC DEVELOPEMENT OF EWOD LAB-ON-A-CHIP FOR CHEMICAL PROCESSOR

P.Dubois[1,2], Dr G. Marchand[1], Dr D. Jary[1], Dr Y. Fouillet[1], Dr M. Vaultier[2]

[1] CEA/LETI, Department of Technology for Biology and Health, Grenoble, France
[2] C.N.R.S, University of Rennes1, Rennes, France

Abstract

Digital microfluidic coupled with EWOD (electrowetting on dielectric) appears very promising for LOC applications, since all the fluidic functions can be achieved by electostatique actuation, simplifying the integration of complex protocols, without pumps or any other moving parts. Basic fluidic functions, such motion, dispensing, merging and mixing are reported here. Biological applications allowed by the combination of these elementary steps are also described. Next, this papers deals with the development of a chemical processor, based on the original combination of ionic liquids (ILs) and open EWOD. The advantages and particularities of ILs used in combination with EWOD are discussed. Finally, a feasibility study on Grieco reaction is presented.

Introduction

In the last decade, lab-on-a-chips (LOC) dedicated to chemical applications have attracted growing interest[1, 2]. Indeed, the miniaturized systems allow reducing the consumption of reagents while optimizing the reactions and yield, producing the desired products faster and in purity. Most of microsystems described in the literature are based on continuous flow. These consist in a network of few micrometer-sized channels embedded into a solid substrate such as glass, silicon, polymers, steel or stainless steel, and ceramics. Within these channels, chemicals are brought together with hydrodynamic pumping[3] or with electro-osmotic flow as reported by S.J. Haswell[4]. Numerous chemical applications have been reported with such technologies, such as Suzuki coupling[5], Wittig reaction[6], Michael addition[7] or Claisen addition[8]. However, continuous flow system suffers from drawbacks, such as frequent channel clogging or difficulties to handle viscous liquids. Ismagilov proposed an digital approach, based on droplets in a channel separated by an immiscible surrounding liquid[9, 10]. However, if this approach allows the creation of new LOC[11, 12], complex protocols are difficult to perform, as individual droplet management is impossible.

An other approach is to handle individual droplets used as microreactors, which are displaced on a two dimension chip[13, 14]. Combinatorial chemistry, multicomponent or multisteps synthesis can be easily performed by the hierarchical mixing of single droplets. To move droplets on a surface several methods such as dielectrophorisis[15], acoustic waves[16] or chemical gradient[17] exist. However, complex protocols require not only displacement but also some basic fluidic functions such as on-chip splitting, dispensing and mixing with a wide range of solutions. The most achieved method to do so is electrowetting on dielectric (EWOD). After a brief presentation of EWOD, this paper deals with the validation of the platform that was developed in our lab. The above cited basic fluidic functions are demonstrated and characterized. These functionalities allow the performing of on chip biological applications. Next, we will focus on the investigation of ionic liquids to perform basic fluidic function and on the use of ionic liquid droplets as microreactors.

EWOD phenomena

The EWOD technology is based on the observation that an electrical field can modify the apparent contact angle of a conducting droplet on a hydrophobic substrate. The shape of a droplet could therefore be reversed by tuning of an electrode (figure 1).

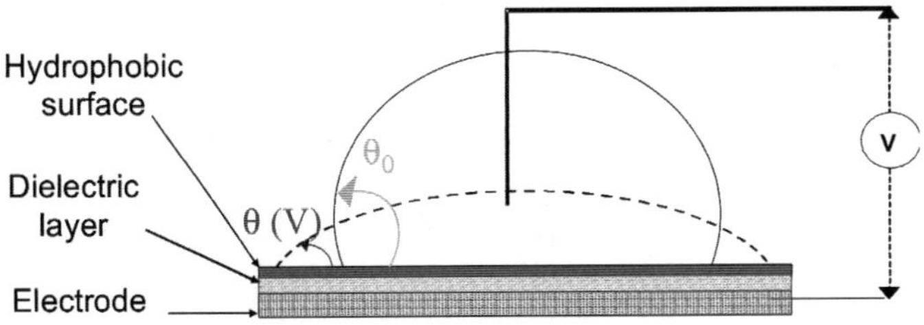

Figure 1: scheme of the droplet deformation under EWOD actuation with a single electrode

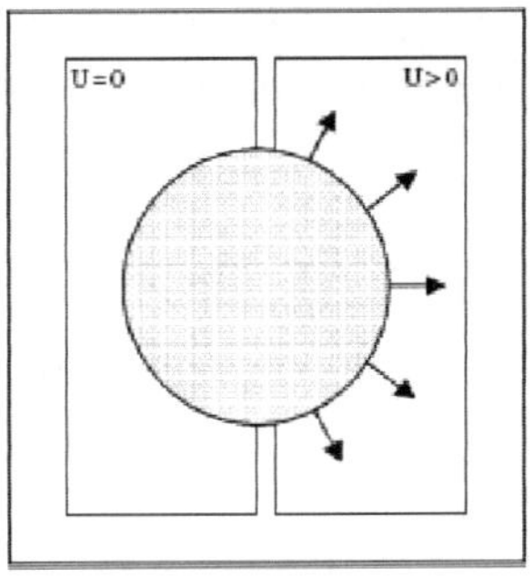

Figure 2: droplet motion under EWOD[19]

A conductive droplet surrounded by a dielectric media such air or oil is placed on an electrode covered by a dielectric layer and a hydrophobic surface. This principle is described by the Lippmann-Young equation (Equation 1) and was first reported by Berge[18]:

$$\cos\theta = \cos\theta_o + \frac{1}{2}\frac{C}{\gamma_{LG}}V^2$$

Equation 1

where θ_o is the natural contact angle, θ is the contact angle when the electric field is actuated, C is the capacitance per unit surface of the layer between the electrodes and the liquid, V is the applied voltage and γ_{LG} is the surface tension between the drop and the surrounding fluid (air or insulating liquid). However, equation 1 is only valid at low voltage. Above a given voltage, a saturation threshold is reached, which corresponds to the minimum contact angle of the droplet. Currently, a total wettability has not been obtained with EWOD. Although the saturation phenomena has not been fully explained yet, it likely seems that it results from high electric fields initiated by the edge effect in the vicinity of the three-phase line[19]

Basic fluidic functions

Motion

For droplet displacement by EWOD actuation, several electrodes are needed, each having a smaller area than the droplet wetting zone. Thanks to a simple system of electrical relays, which makes possible to electrically actuate various electrodes, and therefore to select the zones where electrowetting must act. A droplet placed between two electrodes, one of which is actuated, tends to slide towards the actuated electrode (Figure 2)[20].

Motion can be performed by two technologies: open and closed EWOD. Both are based on micro technology process. A 300 nm gold layer is deposited on oxidized silicon wafer. Electrodes are etched in this gold layer and covered with a 300 nm of silicon nitride used as dielectric layer. Then, the surface is covered with Dupont AF® teflon to get a hydrophobic surface. In open droplet system (ODS – Figure 3), droplets simply lie on electrodes and are in contact with a 30 μm diameter gold wire. This gold wire is used as an electrical catena to apply an actuation between the droplet and electrode. ODS is beneficial from a simplicity and low cost fabrication standpoint[21]. In closed droplet system (CDS – Figure 4), droplets are constrained between electrodes and a conductive top layer. CDS require more sophisticated technological processes than ODS, but allows a better integration, and therefore is more widespread chosen for biological processes.

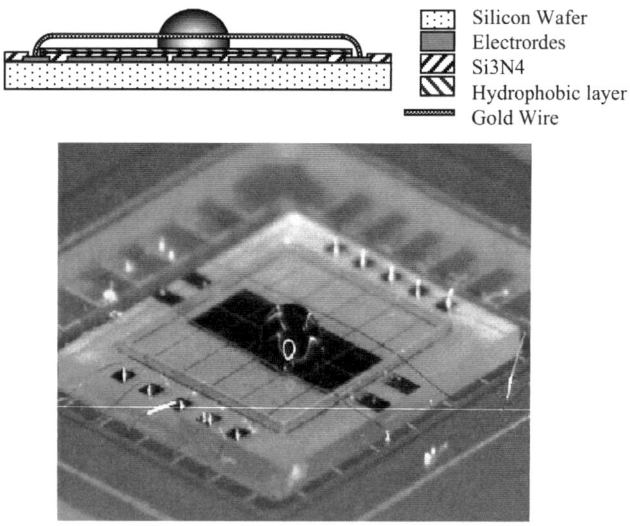

Figure 3: Diagram and picture of open droplet system

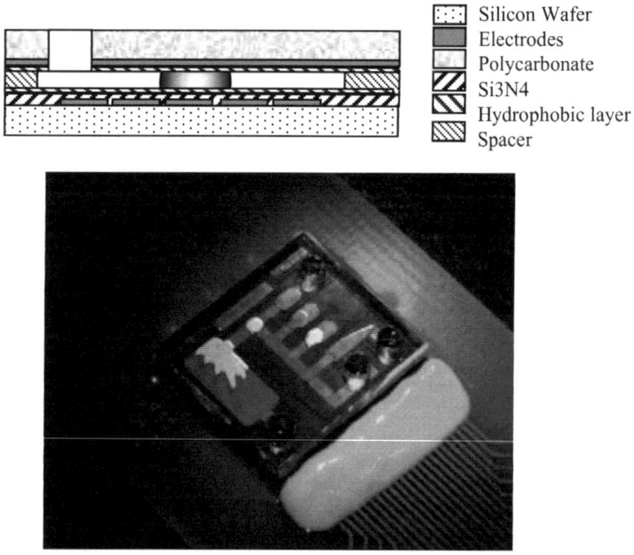

Figure 4: Diagram and picture of closed droplet system

Dispensing

To design LOC, the droplet generation must be integrated and perfectly controlled. EWOD allows the creation of droplet from a reservoir on-chip (figure 5), without needing moving part or pressure[22]. On our CDS chips, sample solutions are deposited on reservoir electrode, through a hole drilled in the top layer. As showed in figure 5, 64 nL droplets are dispensed with a CV below 3% on 800x800 μm electrodes with a 100 μm gap between electrodes and the top layer.

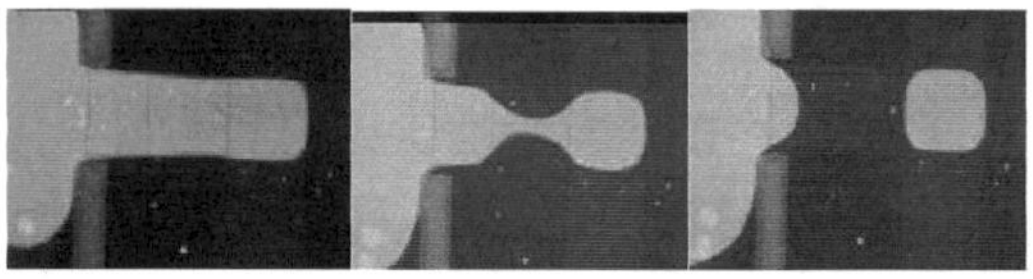

Figure 5: Diagram & Picture of droplet

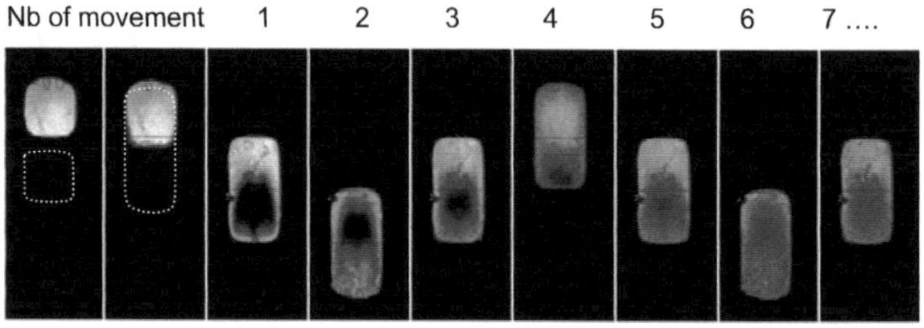

Figure 6: EWOD mixing of water based solution with linear motion

Merging & Mixing

Merging of two droplets can be performed by displacing the droplets towards the same actuated electrode. The mixing of low viscous solution is then ensures by displacement on a row or area of electrodes[23] (figure 6). Our results show that fifteen movements are needed to mix two droplets of 67 nL (95% homogeneous) corresponding to 15 seconds for 1Hz in motion frequency. Quicker mixing (2-3 seconds) can be obtained by increasing the motion frequency.

Biological applications

With the above mentioned basic fluidic functions, we are able to conceive complex protocol. Hence, as an example of biological application, we have worked on a human genotyping based on Polymerase Chain Reaction. A DNA sample solution is dispensed into small droplets on a CDS. Each droplet is then mixed with another droplet containing the specific SNP-PCR mix. On-line detection of genotyping reactions is done using a microscope equipped with a fluorescence detector. Amplification was performed using human placental DNA samples corresponding to 10 DNA copies in 64 nL. Taqman Amplitaq mix (Applied Biosystem) was used for the PCR reagents. The thermo cycling was obtained by placing a Peltier plate under the EWOD chip. The PCR cycles consisted of two steps of 30 seconds: first one at 60°C and second one at 92 °C, with 30 seconds ramping between both. No cross-contamination between positive (drop +) and negative test (drop -) was observed (figure 7-A) and, comparing with PCR performed in standard microtubes of 10 µL, similar cut-off cycles were observed with increases of fluorescence after 30 cycles (figure 7-B). As confirmed by gel electrophoresis analysis (Agilent BioAnalyser 2100), amplification is specific with good yields on CDS[24].

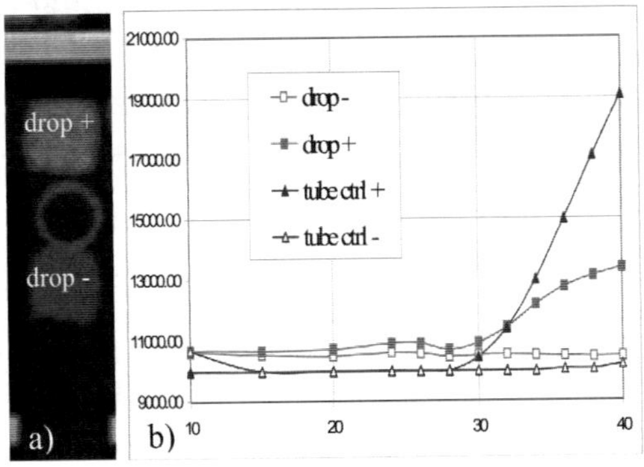

Figure 7: PCR

a) Fluorescence (Fam) of positive and negative droplet after 40 cycles
b) Fluorescence intensity of Fam at various number of cycles – comparison is made to a standard experiment performed in a microtube

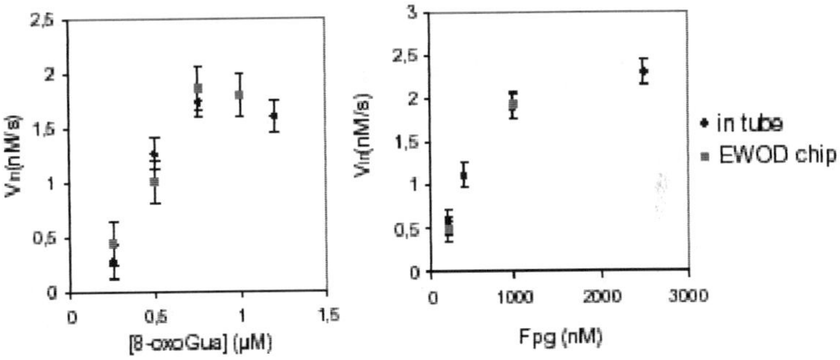

Figure 8: DNA repair

Comparison of initial velocity between on-chip and tube analysis

a) [Fgp] = 1µM / various [8-oxo-Gua]
b) [8-oxo-Gua] = 1,2 µM / various [Fgp]

As a second application, DNA enzymatic repair activities were quantified with modified molecular beacon. Modified molecular beacon consists of small DNA oligonucleotides containing specific DNA lesion and covalently labelled with two fluorophores, which quench each other. When the specific lesion is eliminated, a fluorescent signal appears. On a CDS regulated at 37°C, a fapy glycosylase (Fpg) DNA repair enzyme was tested on a molecular beacon containing an 8-oxo-Guanine oxidative (8-oxo-Gua) lesion, with various concentrations of enzyme and substrate. Tests were performed on 128 nL droplets and compared with 6 µL volume assay on a real time PCR instrument (Stratagene MX 3500P). Initial velocities of enzyme repair activity at constant enzyme or substrate concentrations were identical within experimental error for on chip and in tube reaction (figure 8). A good sensitivity limit of 5 nM of Fpg was found for both methods. This type of DNA repair activity assay on very small samples could be of great interest in the field of cancerology for therapeutic adaptation or type of tumour diagnostic[25]. For both biological applications no loss of enzyme activity due to volume size reduction, high surface to volume ratio or EWOD actuation was detected.

Chemical application

To increase the field of applications, we are currently working on a chemical processor[26]. Most of the applications reported so far are performed in aqueous solutions, which are often not suitable for chemical reactions. Indeed, a chemical processor requires other media, such as volatile organic solvent (VOS). But the use of VOS raises the problem of droplet evaporation. Usually, to prevent such difficulty, droplets are handled in CDS and surrounded with an oil layer. However, the use of organic media under oil could introduce cross-contamination due to the oil/solvent miscibility. We therefore decided to use ILs on our chemical chip, assuming that these particular solvents could be used in open microsystems without evaporating. Recently, ILs have attracted considerable attention. Some of them, the so-called Room Temperature Ionic Liquids (RTILs) are free-flowing liquids at room temperature. RTILs are increasingly used in chemical synthesis as solvents, electrolyte materials and in biological applications due to their remarkable properties such as high ionic conductivity, very low evaporation rate, large potentials range in electrochemistry[27]. Moreover, a unique subclass of ILs, task-specific ionic liquids (TSILs) have greatly widened the utilisation scope of ILs[28-30]. Their functional groups, covalently linked to the cation and/or the anion, confer additional properties to these salts, therefore expanding their potential applications far beyond those of conventional ionic liquids. Moreover, TSILs and RTILs can be combined in solutions; hence, supported synthesis in homogeneous solution can be achieved, which is a major advantage. As ILs are insensitive to evaporation, a top layer or an oil layer are useless, which allows to benefit from the easy and low cost fabrication of ODS. Further, the handling is facilitated, as ODS permits a direct access to droplets.

Ionic liquid & EWOD : static / motion / dispensing / mixing

ILs exhibits clearly different physical properties than water based solution, such as high viscosities (50 to 300 times more than water) and low surface tension (35 to 50 mN.m). Therefore, a study of the comptabilty of ionic liquids with EWOD was necessary. Recent works shows that ILs are sensitive to the EWOD phenomena[31-33] (figure 9). The spreading of 6 ILs on an single electrode has been quantified and compared with one of aqueous solution[34]. As for water based solution, ionic liquids follow the equation 1, quadratic increment in voltage up to a given voltage (saturation voltage), results in the linear increment of the EWOD force '$\gamma(\cos\theta - \cos\theta_0)$'. Above the saturation voltage, a saturation threshold appears.

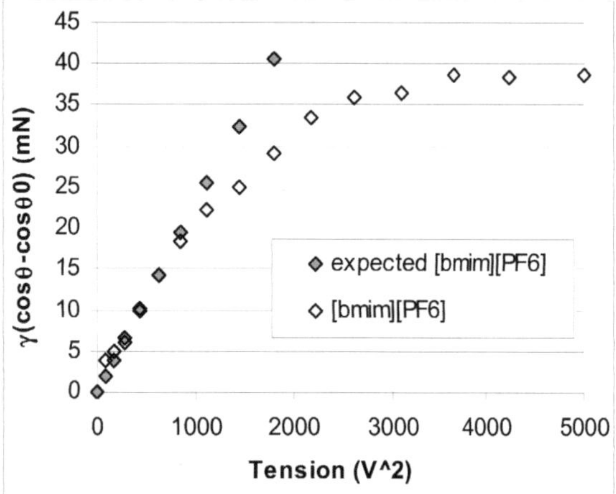

Figure 9: expected electrowetting curves (equation 1) and data raw for [bmim][PF_6]

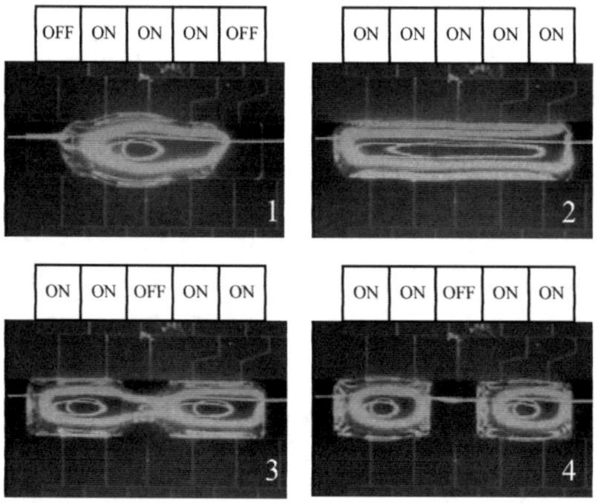

Figure 10: electrodes diagram and pictures of the splitting of 0,4 µL droplet of [tmba][NTf_2] under air

On the one hand, all tested liquids on a same chip exhibit the same saturation voltage. On the other hand, ionic liquids exhibit an electrowetting forces at saturation similar to the one of DI water, but lower than biological buffer.

The movability of all ILs tested was successfully validated. The minimum potential required to move a droplet was measured lower for ILs than for water. We assume that the lower V_{min} are due to a small hysteresis of ILs on the surface and their low surface tension. An other original result linked to hysteresis and surface tension is the splitting of ILs under air. ILs dispensing can be performed by EWOD under air by split of a mother drop, leading to two daughter drops. ILs are the only solution currently known, which allows such splitting under air without top layer (figure 10).

On the other hand, ILs exhibit lower speeds (1-10 mm/s) than DI water (120 mm/s), due to their high viscosity (50 to 300 times more than water). These slow motion and high viscosity raise the problem of mixing. Indeed, slow motion prevent from decreasing the motion frequency and, as showed in figure 11, forth and back flow is fully reversible. Furthemore, the high viscosity slackens dramatically the diffusion process. Thus, it takes approximately 6 min to completely mix two 0,4 μL droplets of [tmba][NTf_2] with a 0,8 Hz motion frequency on a ODS.

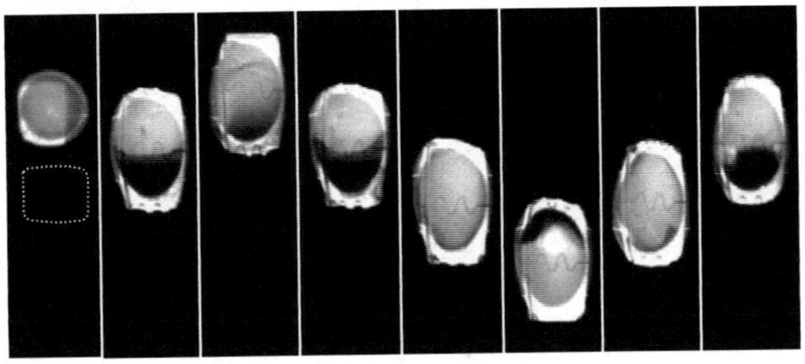

Figure 11: EWOD mixing of [tmba][NTf_2] with linear motion

Grieco reaction

Next to the investigation of these basic fluidic functions with ILs, as a feasibility study, a Grieco's reaction[35] was performed on our chip. Grieco's reaction is an easy, fast and total reaction leading to tetrahydroquinolines and involves four components: an aniline, an aldehyde, an olefin and an acid as a catalyst (figure 12). We used a TSIL, with an aldehyde **1** or an aniline **5** grafted on it. **1** or **5** were reacted with 2 equivalents of different aniline **2** or aldehydes **6**, 10 equivalents of indene **3** and 2 equivalents of trichloroacetic acid. With such conditions at macroscale (in flask), 5 to 10 minutes are needed to get a 100% conversion rate at room temperature (RT). At the microscale, three droplets containing the different reagents were deposited on an ODS and merged together. A 100% conversion rate is reached in 40 minutes in [tmba][NTf_2]. The validation of the droplet reaction was monitored by ESI/MS and liquids chromatography.

The lower rate enhancement in droplet compared to batch is explained by the lack of agitation in droplets. To obtain similar rate enhancement than macroscale at RT, we heated droplets with a resistance underneath. By heating the droplet up to 80°C, the reaction is fully performed in 10 minutes. If it is true that the Grieco reaction takes benefit from the increase in temperature (from a thermodynamical point of view), we assume that the increase in rate enchancement is mainly due to a better mixing. Indeed, with the heating of IL droplet, viscosity decreases significantly and convection flow, such as Marangoni effect, could appears inside the droplet.

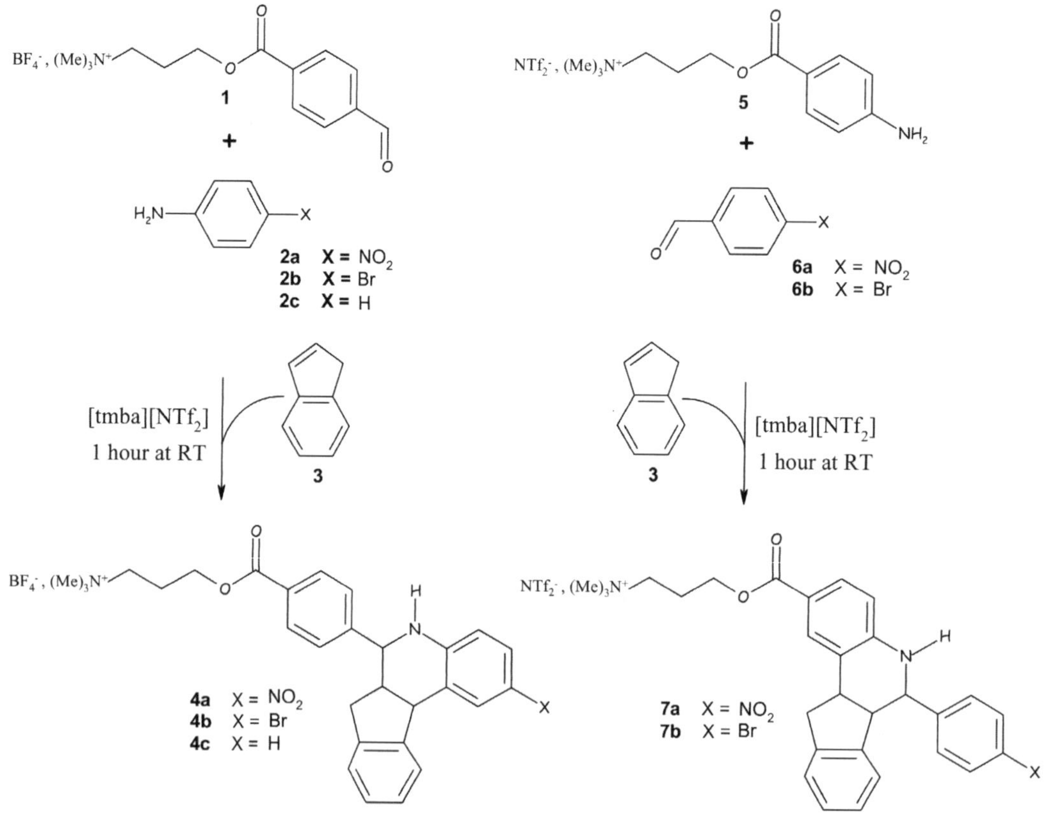

Figure 12: Schematic representation of the Grieco reaction supported on TSIL

To investigate the reason(s) of increase of reaction rate, we are currently working on new detection, such as Raman spectroscopy. Indeed, with such direct detection, we will be able to monitor the Grieco, or whatever reactions, directly inside the droplets.

Conclusion

In summary, digital microfluidic coupled with EWOD actuation appears very promising since all the fluidic functions can be achieved by EWOD actuation, simplifying integration of complex protocols. This is due to the fact that fluidic movements are accomplished electronically without pumps or any other moving parts. Two biological applications have already been achieved, and we are now working on chemical applications.

Ionic liquids are suitable candidates for digital droplet chemistry, as they allow the use of open digital system. Even with their high viscosity and low surface tension they can be move and dispense on a chip. Mixing is a tricky point which is currently under investigations. However, even with this poor mixing, a tri-components chemical reaction was performed inside ILs droplets. Thus, 8 chemical components were synthesized in parallel on our chip.

1. Pennemann, H.; Watts, P.; Haswell, S. J.; Hessel, V.; Lowe, H., *Organic Process Research & Development* **2004**, 8,

(3), 422-439.
2. Dittrich, P. S.; Tachikawe, K.; Manz, A., *Analytical Chemistry* **2006,** Published on web 26 April.
3. Gravesen, P.; Branebjerg, J.; Jensen, O. S., *J. Micromech. Microeng.* **1993,** 3, 168-182.
4. Fletcher, P.; Haswell, S. J.; Paunov, V. N., *Analyst* **1999,** 124, 1273.
5. Greenway, G. M.; Haswell, S. J.; Morgan, D. O.; Skelton, V.; Styring, P., *Sensors and Actuators B: Chemical* **2000,** 63, (3), 153-158.
6. Skelton, V.; Greenway, G. M.; Haswell, S. J.; Styring, P.; Morgan, D. O.; Warrington, B. H.; Wong, S. Y. F., *Analyst* **2001,** 126, 7.
7. Wiles, C.; Watts, P.; Haswell, S. J.; Pombo-Villar, E., *Lab on a Chip* **2002**, (2), 62-64.
8. Mu, J. X.; Yin, X. F.; Wang, Y. G., *Synlett* **2005**, (20), 3163-3165.
9. Zheng, B.; Ismagilov, R. F., *Angewandte Chemie International Edition* **2005,** 44, (17), 2520-2523.
10. Nisisako, T.; Torii, T.; Higuchi, T., *Lab on a Chip - Miniaturisation for Chemistry and Biology* **2002,** 2, (1), 24-26.
11. Shestopalov, I.; Tice, J. D.; Ismagilov, R. F., *Lab on a Chip - Miniaturisation for Chemistry and Biology* **2004,** 4, (4), 316-321.
12. Chan, E. M.; Alivisatos, A. P.; Mathies, R. A., *Journal of the American Chemical Society* **2005,** 127, (40), 13854-13861.
13. Taniguchi, T.; Torii, T.; Higuchi, T., *Lab on a Chip* **2002,** 2, 19-23.
14. Velev, O.; Prevo, B. G.; Bhatt, K. H., *Nature* **2003,** 426, 515-516.
15. Jones, T. B.; Gunji, M.; Washizu, M.; Feldman, M. J., *Journal of Applied Physics* **2001,** 89, (2), 1441-1448.
16. Guttenberg, Z.; Muller, H.; Habermuller, H.; Geisbauer, A.; Pipper, J.; Felbel, J.; Kielpinski, M.; Scriba, J.; Wixforth, A., *Lab on a Chip* **2005,** 5, (3), 308-317.
17. Yamada, R.; Tada, H., *Langmuir* **2005,** 21, (10), 4254-4256.
18. Quilliet, C.; Berge, B., *Current Opinion in Colloid & Interface Science* **2001,** 6, (1), 34-39.
19. Mugele, F.; Baret, J.-C., *Journal of Physics Condensed Matter* **2005,** 17, 705-774.
20. Bavière, R.; Boutet, J.; Fouillet , Y. SHF-Microfluidics, 12-14 December 2006, Toulouse.
21. Fouillet, Y.; Jeanson, H.; Chartier, I.; Buguin, A.; Silberzan, P., *Houille Blanche* **2003,** 37, (4).
22. Cho, S.-K.; Moon, H.; Kim, C.-J., *Journal of Microelectromechanical systems* **2003,** 12, 70-80.
23. Paik, P. Y.; Pamula, V. K.; Fair, R. B., *Lab on a Chip* **2003,** 3, 253-259.
24. Fouillet , Y.; Jary, D.; Brachet, A. G.; Berthier, J.; Blervaque, R. D., L; Roux, J. M.; Achard, J. L.; Peponnet, C. Fourth International Conference on Nanochannels, Microchannels and Minichannels, June 19-21, 2006; Limerick, Ireland.
25. Jary, D.; Chollat-Namy, A.; Boutet, J.; Chabrol, C.; Castellan, G.; Gasparutto, G. P., C Nanotech, 2006; Boston, pp 554-557.
26. Fouillet , Y.; Jary, D.; Brachet, A. G.; Chabrol, C.; Boutet, J.; Clementz, P.; Lauro, D.; Charles, R.; Peponnet, C. Micro Total Analysis System, 2005; Boston, pp 58-60.
27. Welton, T., *Chemical Review* **1999,** 99, 2071-2083.
28. Vaultier, M.; Gmouh, S.; Hassine, F. FR 2857360, Application n°2003-8413, WO**2005**005345.
29. Lee, S.-g., *Chemical Communications* **2006**, 1049-1063.
30. Davis Jr., J. H., *Chemistry Letters* **2004,** 33, (9), 1072-1077.
31. Millefiorini, S.; Tkaczyk, A. H.; Sedev, R.; Efthimiadis, J.; Ralston, J., *Journal of American Chemical Society* **2006,** 128, (9), 3098-3101.
32. Dubois, P.; Marchand, G.; Fouillet , Y.; Peponnet, C.; Chabrol, C.; Berthier, J.; Vaultier, M. Micro Total Analysis System, 2005; Boston, pp 412-414.
33. Chatterjee, D.; Hetayothin, B.; Wheeler, A. R.; King, D. J.; Garell, R. L., *Lab on a Chip* **2006,** 6, 199-206.
34. Dubois, P.; Marchand, G.; Fouillet , Y.; Berthier, J.; Douki, T.; Hassine, F.; Gmouh, S.; Vaultier, M., *Anal. Chem.* **2006,** 78, (14), 4909-4917.
35. Grieco, P. A.; Bahsas, A., *Tetrahedron Letters* **1988,** 29, (46), 5855-5858.

A PROPANE-AIR MICRO JET PREMIXER

Jian Zhang, Louisiana State University, Baton Rouge, LA
Kevin W. Kelly, Louisiana State University, Baton Rouge, LA
Sumanta Acharya, Louisiana State University, Baton Rouge, LA
Charles Becnel, Mezzo Technologies, Baton Rouge, LA

Introduction

Mezzo Technologies fabricates micro tube heat exchanger panels that provide extremely high rates of heat transfer per unit volume with very little pressure drop penalty[1,2]. These heat exchanger panels can then be assembled in a variety of ways to meet the needs of a specific application. The heat transfer performance advantages associated with these panels are the predicted result of decreased channel size. These panels, originally conceived for heat exchanger applications, are now being investigated for use as mixers and reactors[3]. Modified versions of these panels essentially provide sophisticated plumbing that enables mixing on the micro scale of the bulk flows of two (or more) gas, liquid or two-phase constituents. The mixing process occurs quickly, within a very compact volume, and in most cases does not involve substantial pressure drops of any of the constituents being mixed. Furthermore, it is straightforward to combine micro scale heat exchangers with these mixer/reactors to simultaneously provide rapid, uniform mixing *and* temperature control. In this paper, a propane-air premixer for a gas turbine application is described which consists essentially of a single, modified heat exchanger panel. While the performance of premixer has not yet been completely characterized, initial results look promising

The motivation to build an improved propane-air mixer is based on the fact that the power generation industry is faced with increasingly stringent emissions requirements for ozone precursors, such as nitrogen oxides and carbon monoxide. To achieve lower pollutant emissions, gas turbine manufacturers have adopted lean premixed (LP) combustion as a standard technique, particularly for natural gas applications. LP combustion achieves low levels of pollutant emissions without additional hardware for steam injection or selective catalytic reduction. By premixing the fuel and air well, localized regions of near stoichiometric fuel-air mixtures are avoided and a subsequent reduction in thermal NOx can be realized[4].

Most gas turbine fuel-air mixers are based on "single" point injection, where the fuel is shot into a swirling air stream from a "single" injection point (which in fact is a close-packed array of injection holes)[5]. A new multi-point injection premixer that injects the fuel into the air via uniformly dispersed micro jets across the entire cross section of the flow field has been fabricated and tested. Relative to "single-point" fuel injection, multi-point fuel injection offers numerous advantages, such as significantly shorter mixing length and time scales, as well as a very simple, compact geometry.

The propane-air premixer that has been built and tested is similar in form to the heat exchanger panels shown in Figure 1[6]. Figure 1(a) provides a view of one side of a panel. These panels can have arbitrary footprint (width x length) measured in values of tens of square inches, and have a thickness on the order of one eighth of an inch. Each panel is a hollow, monolithic shell perforated with hundreds to tens of thousands of micro-tubes. Flow of one fluid passes through the panel in a direction perpendicular to the plane of the panel via a multitude of micro tubes. The second fluid enters the slot-shaped opening at the end of the panel, weaves through a bank of short micro-tubes, and exits the other end of the panel. Figure (b) shows the internal structure of the panel (a field of short tubes) shown in Figure (a).

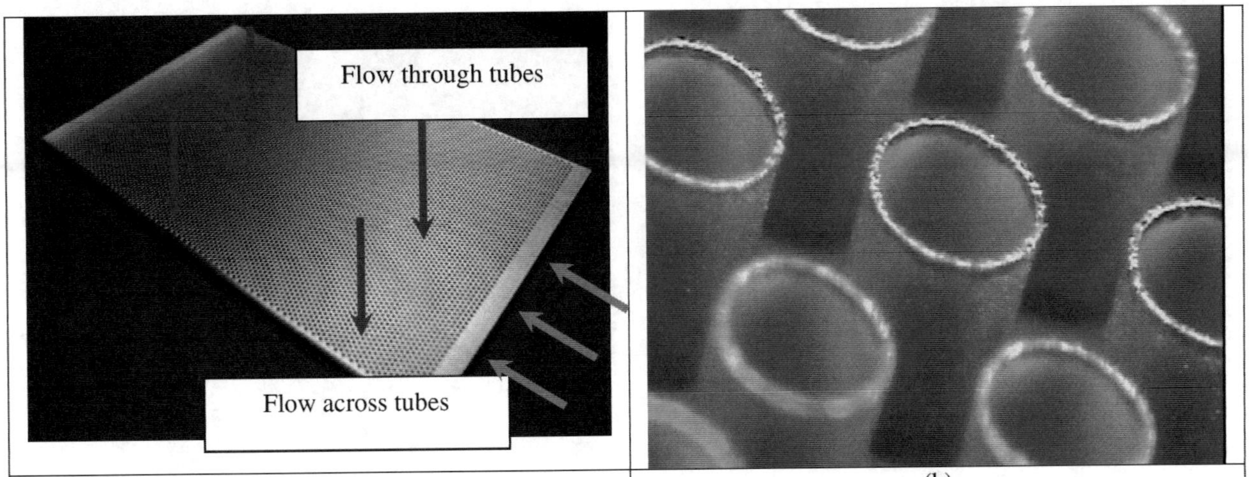

(a) (b)

Figure 1 – (a) Micro-tube heat exchanger panel illustrating the flow paths and (b) internal geometry of the micro-tube panel[1].

The heat exchanger panels can be arranged into a variety of configurations to meet the demands of a particular heat exchanger application. For air/liquid applications, the heat exchanger panels are stacked, aligned, and joined together to form a "laminated" micro-tube heat exchanger as shown Figure 2.

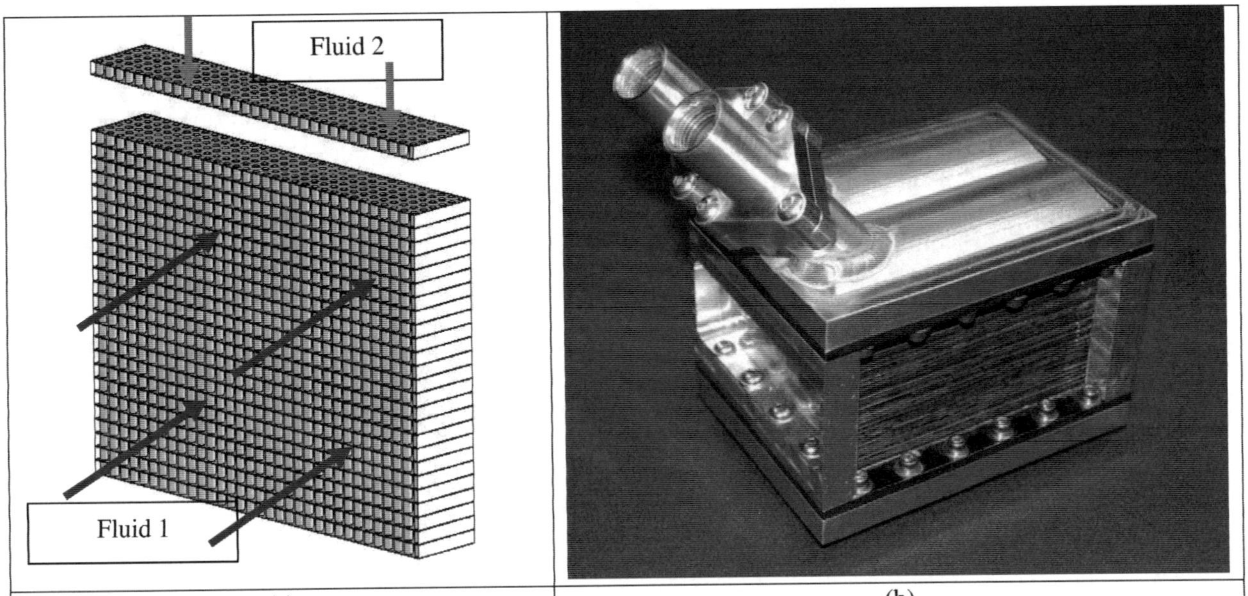

(a) (b)

Figure 2 – (a) Schematic of laminated micro-tube heat exchanger concept and (b) photograph of subscale RFAC heat exchanger utilizing the laminated concept.

Another possible arrangement of panels is shown in Figure 3. In this arrangement, multiple panels are assembled into an inline array as shown in Figure 3 The panels are separated by a small gap (on the order of 1mm) which trips the boundary layer in the fluid flowing through the holes, increasing the heat transfer with a small, but acceptable, pressure drop penalty.

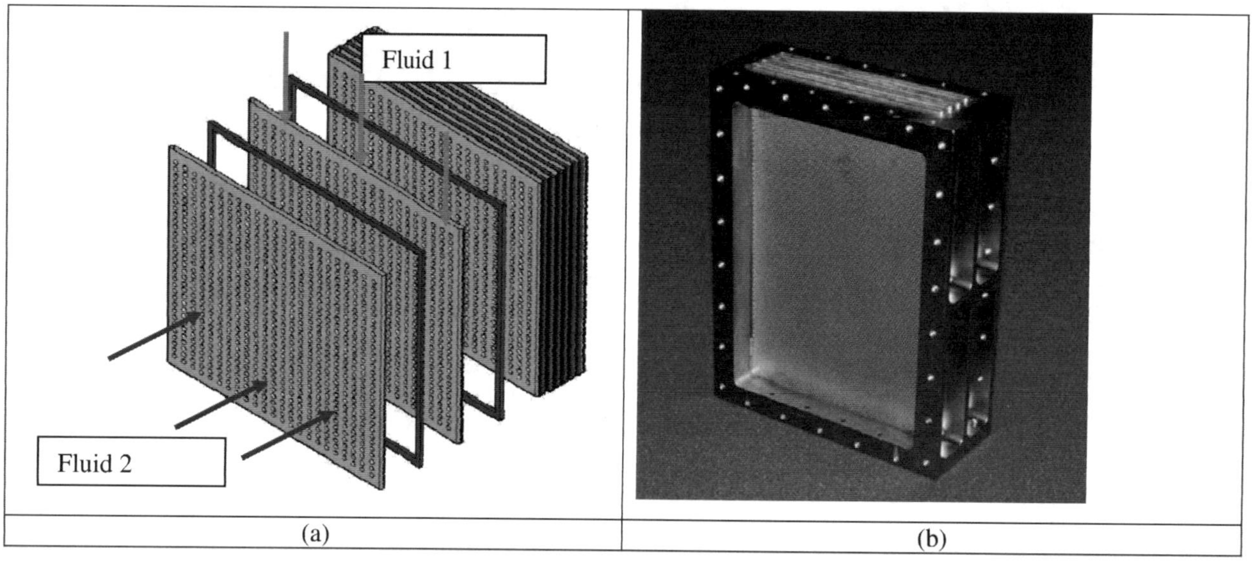

Figure 3 – (a) Illustration of inline array of panels with small gaps between panels and (b) sub-scale liquid heat exchanger prototype.

Mezzo Technologies later modified the heat exchanger panels shown in Figure 1 to produce the Micro Jet Cooling Array (MJCA), a product designed to cool surfaces with high heat flux (such as high power electronics, laser diode arrays, etc.)[6]. The device uses an array of impinging microjets, with diameters on the order of a few hundred micrometers, to cool a surface. Each micro jet establishes over a very small area a region of high heat transfer normally associated with impinging jets. The MJCA differs from a standard array of impinging jets in important ways. In a standard array configuration where a large number of jets impinge on the target, the coolant is collected at the periphery of the target. The performance of this design will be poor for two important reasons: First, each jet is influenced by cross flow of the coolant from the neighboring jets. This cross flow can wash out the impingement effect of jets not at the center of the target. Second, if the exhaust flow is collected at the periphery of the target, a substantial pressure drop may exist between the center of the target and its periphery. This could then result in significant variation in jet velocity across the array, with most of the flow occurring through the jets at the periphery of the impinging jet array.

The MJCA is a 3-dimensional structure that provides an array of microjet channels and a fluid removal system that effectively isolates neighboring jets from the deleterious effect of jet-to-jet cross-flow. A schematic of the underlying principle and photographs of MJCA are shown in Figure 4. The MJCA has a honey-comb structure identical in concept to the cross flow heat exchanger shown in Figure 1, with a typical thickness of a few millimeters and lateral lengths of the order of centimeters. The honey-comb's small-diameter channels act as the microjet nozzle array. The coolant flows through the nozzle array and impinges on the target as an array of high velocity submerged jets. The coolant is then locally exhausted, through a number of ports surrounding each microjet nozzle. *These return holes represent the only substantive difference between the MJCA and the cross flow heat exchanger shown in Figure 1.* The exhausted coolant flows within the hollow honey-comb structure towards the periphery of the MJCA where it is collected and removed. The localized exhausting of each microjet greatly reduces cross-flow between neighboring jets. Thus, the performance of a single submerged microjet can be replicated over a very large area.

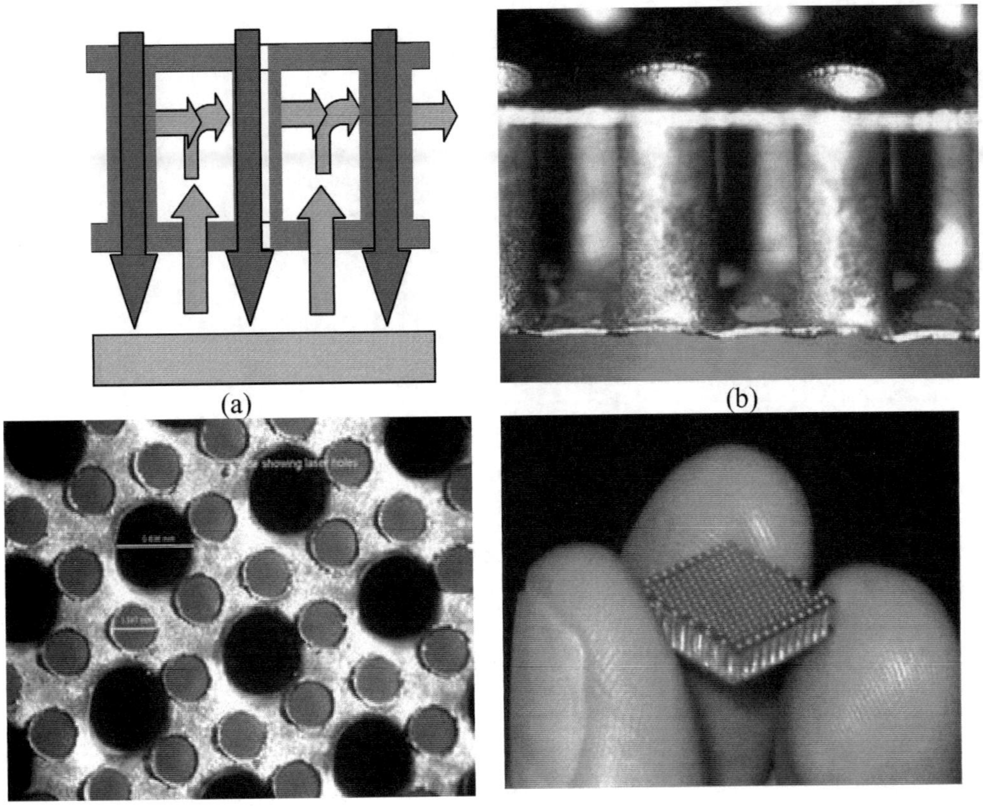

(c) (d)

Figure 4 (a) Schematic of the MJCA concept. (b) Side view of the fabricated MJCA. The conduit connecting top and lower surface are the jet inflow conduits. The holes in the lower surface are the exhaust ports, (c) Bottom view of the MJCA with the jet impingement holes (large holes) surrounded by the exhaust ports (small holes), (d) A fabricated 1x1 sq. cm MJCA[2].

The micro injection propane-air premixer which is described is virtually identical in structure to the MJCA. The gaseous fuel (propane) is fed into the premixer through the in-plane channels of the panel. Holes on one side of the panel provide the means by which the fuel escapes from within the panel, forming a distributed array of micro jets. Air flows in a direction perpendicular to the plane of the panel through the array of holes in the panel. Depending upon direction of the array of fuel jets, the fuel is mixed with the air in either a distributed co-flow or counter-flow configuration. The counter-flow configuration is shown in Figure 5.

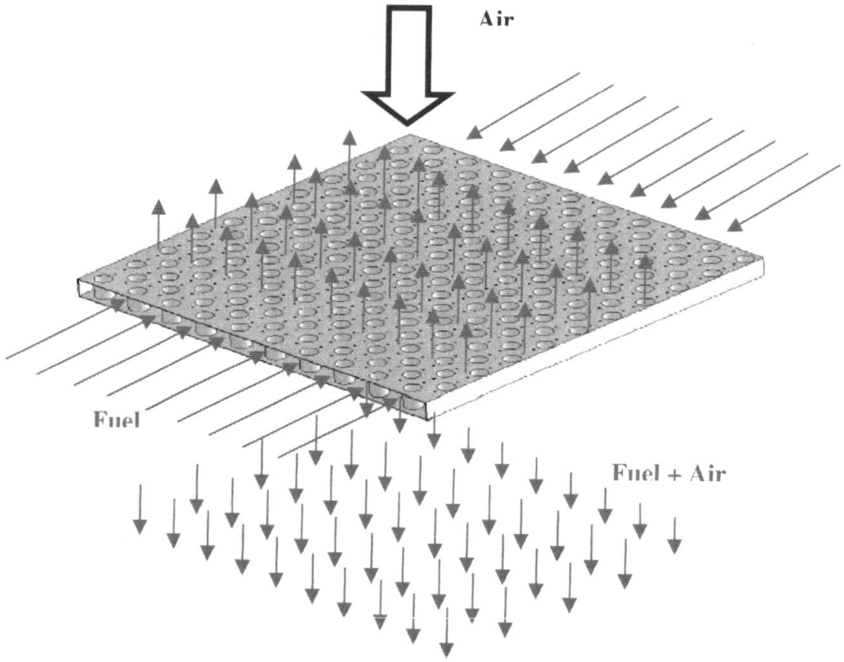

Figure 5: Illustration of mixing of the premixer

A micro air-fuel premixer for gas turbine engines that utilizes micro manufacturing approaches developed at Louisiana State University[7] and being commercialized by Mezzo Technologies was designed and is in the process of being evaluated. The purpose of the test program thus far completed is to determine if the premixer provides better mixing and, as a result, the ability to maintain steady combustion at leaner air-fuel ratios and reduced NOx. To evaluate the new premixer concept, the tasks listed below were completed. This paper provides results associated with some of those tasks including a description of the premixers manufactured, the experimental apparatus, and some of the test results that were collected.

1. The micro injection premixer was designed and fabricated. An apparatus to test the injection premixer was also fabricated.
2. Fundamental experiments were completed to quantify the pressure drop of fuel through the premixer as functions of flow rate. In addition, experiments were completed to characterize the variation of jet velocity across a premixer.
3. With the premixer installed in a combustor, experiment data was obtained for the combustor that indirectly evaluated the performance of the premixer. Data included heat release rate, NOx concentration and the equivalence ratio at lean blow out (LBO).
4. These experimental results were compared to results in the same combustor where the air and fuel were premixed far upstream of the combustion region.

Description of Premixer and Experimental Apparatus

A. Apparatus Description

Schematic of the two configurations of the experimental apparatus are shown in Figure 6a and 6b. Figure 6a describes the scenario where the micro jet premixer is *not* installed. A swirl chamber is located one meter upstream of the premixer. Measured mass flow rates of air and fuel are injected into the swirl chamber. The air-fuel mixture then passes through a 2.54 cm (one inch) diameter pipe of length one meter before suddenly expanding into a quartz tube with diameter 5.5 cm. The plane where the expansion occurs serves as the dump plane where the flame is held during combustion experiments. This case provides a baseline for understanding the combustion performance of this combustor using thoroughly premixed air-propane in the absence of any effects associated with the premixer.

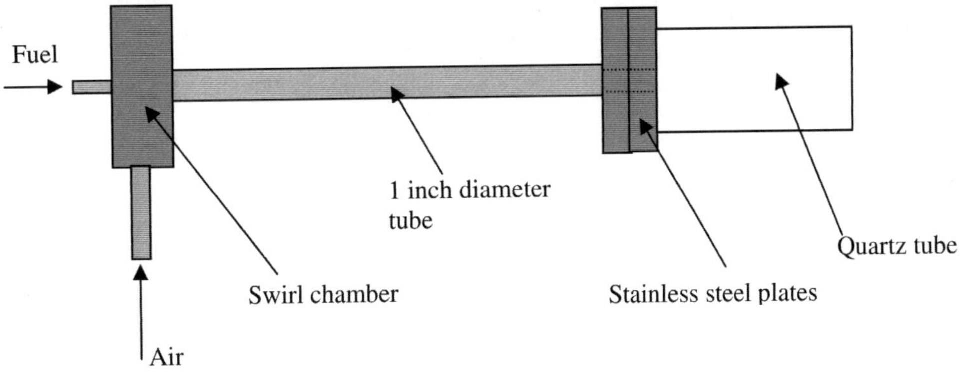

Figure 6a: Apparatus without micro jet premixer installed

Figure 6b describes the experimental apparatus for the cases where the premixer is installed. Geometrically, the length of the apparatus is virtually identical to Figure 6a except that the premixer, with width 3.2 mm is sandwiched between to fixture plates, making the total length of the apparatus 3.2 mm greater. Two fuel injection options are available in Figure 6b. The first option allows the fuel to be injected into the swirl chamber where it is mixed far upstream of the dump plane with air. This case is identical to the case described in Figure 6a (without the premixer installed), except that the premixer *is* physically installed in the flow stream. The second option allows fuel to be injected via the micro jet premixer in a counter flow orientation just upstream of the dump plane.

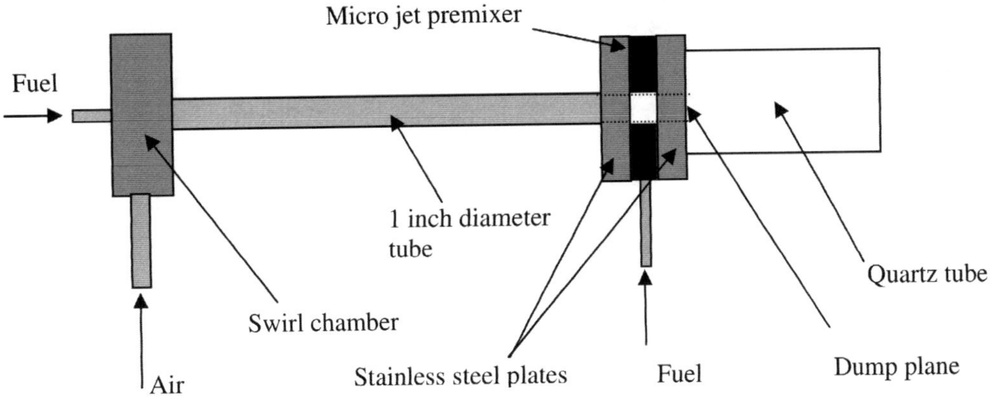

Figure 6b: Apparatus with premixer installed

The combustion air is supplied from a large storage tanks fed by a 290 psig 450ACFM AtlasCopco air compressor. Before the air is fed into the combustor, it is regulated to 160 psig by a pressure regulator and is filtered by a filter section. An air flow meter is used to monitor the volumetric flow rate of the air. A needle valve is installed upstream of the air flow meter to adjust the air flow rate. A pressure gage is installed immediately downstream of the flow meter to monitor the air pressure exiting the flow meter. The air from the exit of the flow meter is delivered into the swirlchamber through four inlet holes. The range of flow rates used for experiment is from 10CFM to 20CFM.

The fuel (propane) used for combustion experiment is contained in a 25 gallon (95 liter) propane tank. The pressure of the fuel is regulated by a compressed gas pressure regulator to 80 psi. The volumetric flow rate of fuel is controlled and monitored by a valve-controlled flow meter. A flashback arrester and check valve (AGAFR50) is installed in the fuel line with the aim of protecting the propane tank from an accidental return of flame. A pressure gage is installed in the fuel line downstream of the fuel flow meter to monitor the pressure drop of fuel in the combustor. The fuel is delivered either into the swirl chamber or into micro injection premixer.

B. Premixer description

One of four propane-air premixers that were fabricated and tested are shown in Figure 7a and 7b. Parameters that differentiate each premixer design include air channel spacing, and fuel jet diameter and spacing. To produce each premixer, headers were attached to open ends of panels similar to that shown in Figure 1. These headers provide access for propane into the internal cavities within each premixer panel. An array of micro holes was drilled into one face of the premixer panel over a one inch diameter region in the center of the panel. Gaskets were attached to both faces of the premixer panel. One inch diameter openings in each gasket were aligned with the region containing the micro jets. The gasket-covered surfaces were sandwiched between two stainless steel plates (shown in Figure 6) such that the opening in the gaskets aligned with the one inch tube upstream. The thickness of all injection plates fabricated was 3.2mm. The diameter of the air channels was 1.8mm diameter. Two different air channel geometries were examined: staggered and square. For the case of the square array, the center-to-center distance between two neighboring air channels is 2.6 mm. The staggered array was equilateral with all neighboring air channels 2.6 mm apart.

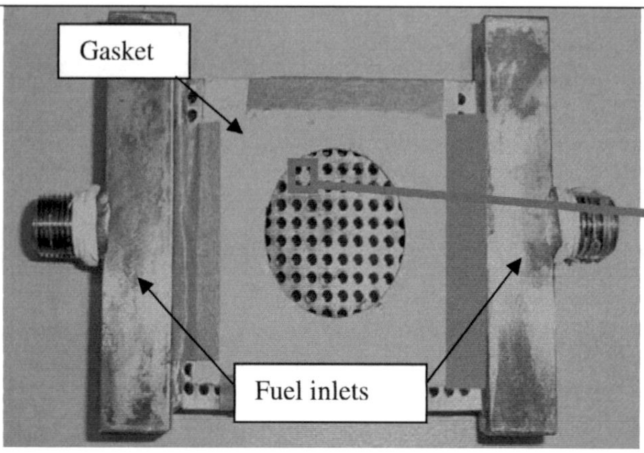

Figure 7a: Premixer panel covered with gaskets

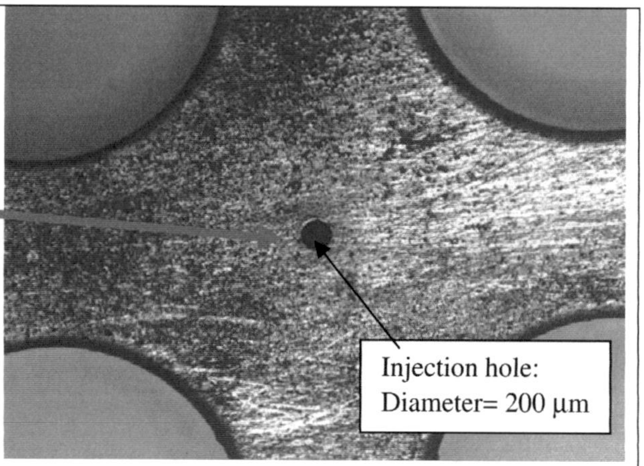

Figure 7b: Magnified view of propane jet

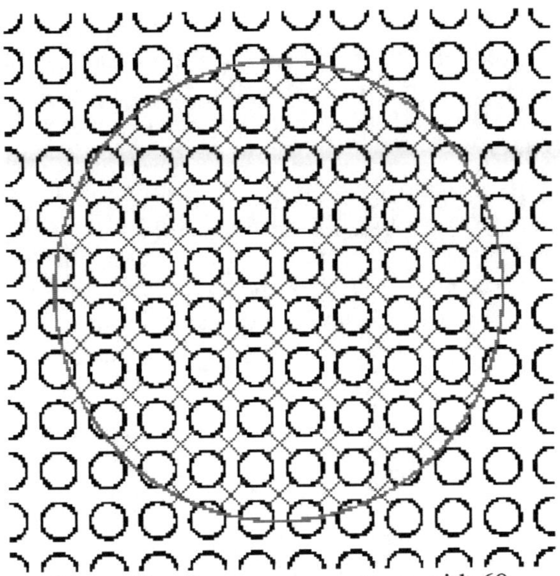

Figure 7a: Square array pattern with 69 injection holes

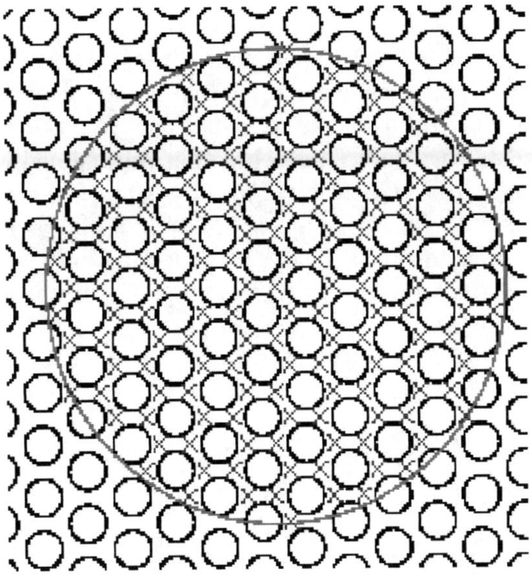

Figure 7b: Staggered pattern with 148 injection holes

The first two premixers used the square (aligned) air channel array shown in Figure 7a. Both of these cases utilized 69 propane jets (locations shown with X). In one case (A-150-69), the diameter of the holes was 150 micrometers, in the other (A-200-69) the diameter was 200 micrometers. The second two premixer plates involved the staggered array of air channels shown in Figure 7b. Again, the number of holes (locations shown with X) for both of these premixers was the same, and the micro jet hole diameter was either 150 or 200 micrometers (S-150-148 and S-200-148). As evident from Table 1, the four cases provide approximately a four-fold range in total area for fuel micro jets. For a given mass flow rate of air and a given equivalence ratio (fuel/air ratio), the velocity exiting the micro jets will be inversely proportional to the total micro jet area. Therefore, these different cases make it possible to determine the importance of momentum ratio with respect to mixing.

Table 1: Parameters for the injection plates

Plate	Pattern	Injection hole size (μm)	Injection hole number	Total area of fuel holes (mm^2)	Total area of air channel (mm^2)
A-150-69	Aligned	150	69	1.22	164.30
A-200-69	Aligned	200	69	2.17	164.30
S-200-69	Staggered	200	148	4.65	189.72
S-150-148	Staggered	150	148	2.61	189.72

Experiments and Results

A. Micro jet velocity measurements

To quantify the performance of the premixer, tests were performed to measure jet velocity variation across the array of micro jets. The ability to efficiently achieve a uniform air-fuel mixture is obviously enhanced if the velocity (and mass flow rate) from each jet is essentially identical. If the distribution is not uniform (i. e. there exits substantial variations in jet velocity across the premixer), then the performance of the premixer would be expected to be greatly compromised.

Premixer A-200-69 (a square array) was used in this experiment. A hot film probe, mounted on a traverse stage was located 1mm in front of the plane containing the micro jet holes. In this experiment, air, instead of propane flowed through the premixer and exited the micro jet holes. Velocity measurements were taken along both the vertical and horizontal lines shown in Figure 8a below. In both the horizontal and vertical direction, the step size between measurements was 125μm, using a Velmex traverse system. The cross marks shown on Figure 8a mark positions of jet holes. The results of the horizontal scan (the vertical scan was identical) are shown in Figure 8b. Along either scan direction, the probe passes over a total of nine jets ranging from positions near the periphery of the jet array to the very center. Variations in the jet velocity profile are minimal, with the differences more likely the result of measurement location varying with respect to each hole than actual differences between flow rate. This test demonstrated that the injection of gaseous fuel from each jet is very uniform.

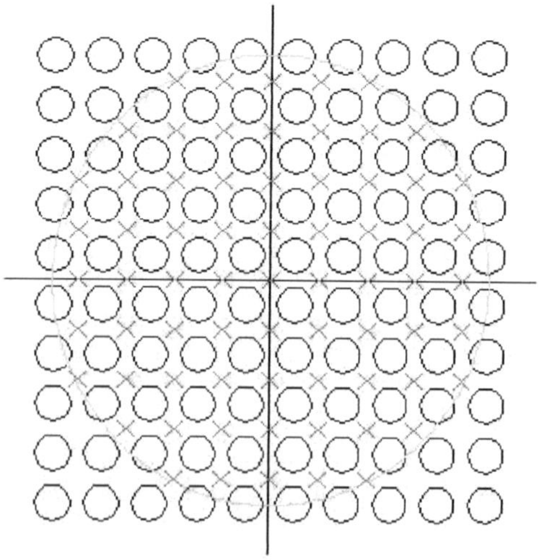

Figure 8a: Horizontal and vertical lines over which velocity measurements taken.

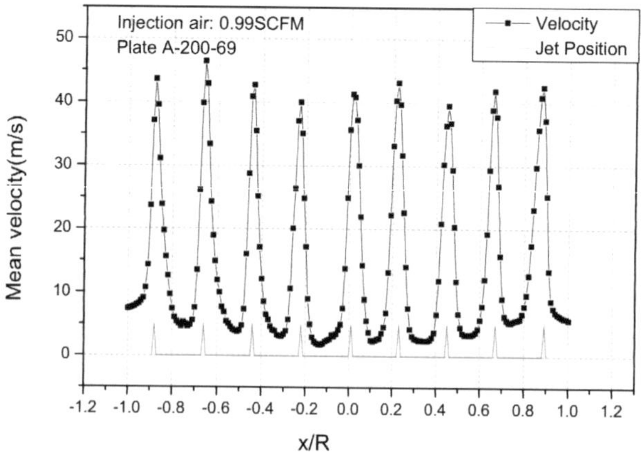

Figure 8b: Mean velocity measurements over horizontal scan.

B. Equivalence Ratio at Lean Blow Out

The equivalence ratio at lean blow out (LBO) for a given combustor defines the minimum fuel/air ratio at which steady combustion can be achieved. Normally, a lower lean blow out limit is pursued because it allows combustors/burners to be able to run continuously at a lower fuel equivalence ratio to achieve a lower NOx concentration emission. The lean extinction limit mainly depends on the velocity distribution in the combustion chamber (which involves the geometry of the combustor), fuel type, and the molecular mixing of the fuel and the air.

The equivalence ratio at LBO was measured for five cases. The cases are: i) counter injection of the array of micro jets into oncoming air using plate A-200-69; ii) counter injection of the array of micro jets into oncoming air using plate S-200-148; iii) counter injection of the array of micro jets into oncoming air using plate S-150-148; iv) air and fuel premixed far upstream of the dump plane with the premixer plate not installed; and, v) air and fuel premixed far upstream of the dump plane, but with the premixer installed and not functioning in any capacity other than as a screen. Theoretically, the cases where air and fuel are mixed upstream in the swirl chamber should provide comparison of the micro jet premixer plates against well premixed cases.

The results are shown in Figure 9. The cases involving injection of fuel via the micro injection plate have the lowest lean extinction limits compared with the other two combustion cases. Therefore, the improvement of lean extinction limit for the counter injection case are believed to be the result of better mixing of fuel compared with the case with the dummy injection plate.

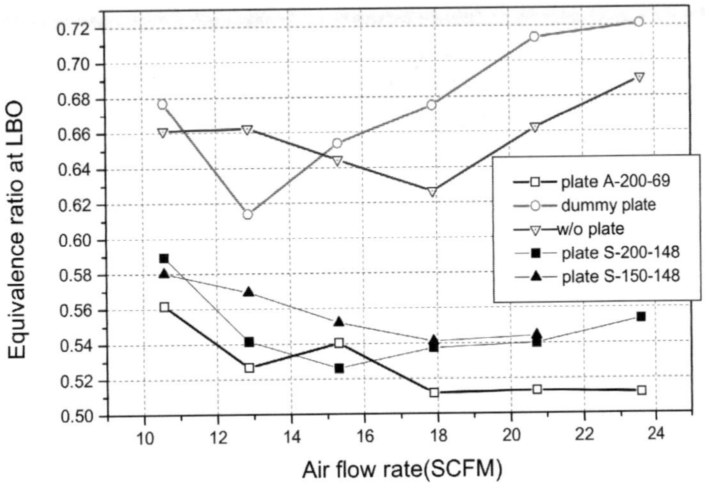

Figure 9: Equivalence Ratio at LBO

C. Heat Release Rate

CH chemiluminescence is a good indicator of the heat release (reaction) rate from a combustion reaction. The fuel and air reaction rate is enhanced when the air and fuel are well mixed at a molecular level. CH luminescence signal was measured during combustion experiments for each of the five cases that were described in the previous section. The CH signal was captured using a silicon Melles-Griot photodiode with a 430 nm centered-optical filter is used to measure the integral of CH signal. The photodiode was fixed 30cm in front of the combustion chamber. During the experiment, natural light was blocked to decrease the noise for CH light measurement.

Typical results are shown in Figure 10. These results are summarized as follows:

- With the counter injection micro mixers, the CH light signal is higher than cases not involving injection of fuel via micro jets.
- Comparing premixers S-200-148, S-150-148 and A-200-69, the CH signal monotonically increases as the injection velocity of the fuel increases (or as the total micro jet cross section area decreases). This implies that higher momentum jets produce better mixing than low momentum jets.
- These results were duplicated across a range of air flow rates (from 11 CFM to 28 CFM)

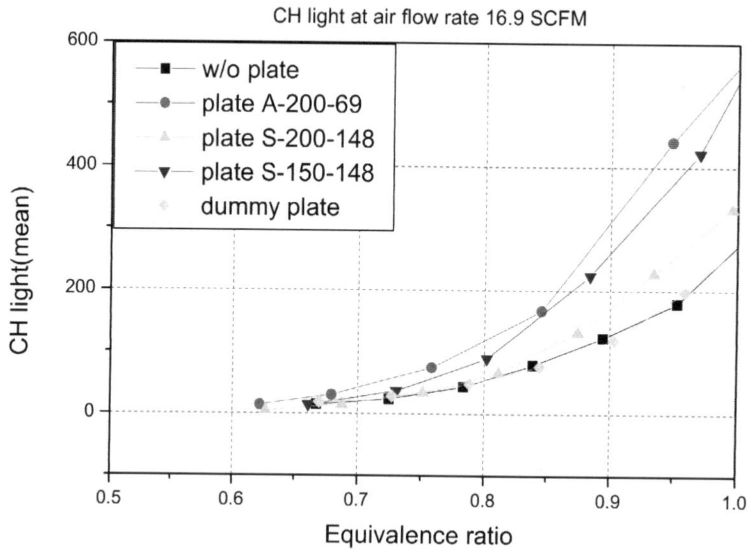

Figure 10: Heat release rate at air flow rate16.9 SCFM

D. NOx measurements

Concentration of NOx at the exit plane of the combustor was measured to investigate the influence of different mixing methods associated with cases 1-5 (previously described). The exhaust gas samples were collected with a stainless-steel suction probe and transferred through an electrically Heated Transfer Sample Line (TSL-H) to a fast chemiluminescence detector (Cambustion *f* NOx400). The analyzer was calibrated using 100ppm NOx gas and 0 ppm pure nitrogen gas by using a DCS-400 Dynamic Calibration System providing accuracy better than 1 % of full scale, with negligible deviations. The response time of the analyzer is around 4ms, which makes the instrument suitable to obtain transient NOx values. Regular calibration checks were performed to eliminate errors associated with drift. The region of primary interest in these experiments was the lean regime of combustion. The variables that are most important with respect to NOx formation are listed below[8]:

- The combustion temperature has the most important effect.
- The combustion pressure only has a neutral or slightly negative effect on the NOx formation.
- The fuel type has some effect, especially at the temperature above 1700K.
- Nonuniformity of fuel/air mixing, has an important effect on NOx emission[9]. Over a range of lean air-fuel conditions, the NOx emission concentration increases as nonuniformity of the local fuel/air equivalence ratio increases. Interestingly, at the equivalence ratio slightly lean of stoichiometric, the average NOx emission decreases with increasing the nonuniformity of the air-fuel mixture.

All the combustion experiments were performed at atmosphere pressure without preheating, using propane as the fuel. Since pressure and inlet temperatures were constant, air flow rate, equivalence ratio (which directly relates to the theoretical combustion product temperature) and the uniformity of the mixing of the fuel and air are the primary factors controlling NOx formation. The suction probe used to obtain gas samples to measure NOx concentration was located at the center of the combustor exit.

Figures 11 and 12 provide experimental results over the lean regime at the two extremes of air flow tested. Figure 11 provides results at the air flow rate 11.2 SCFM. At this flow rate it was found that the NOx emissions were lower for the cases with injection plate A-200-69 and plate S-150-148 (the cases with the higher injection velocity for a given fuel mass flow rate). A possible reason is that micro injection plates A-200-69 and S-150-148 generate better mixing that results in the fewer regions of near-stoichiometric fuel-air, compared with plate S-200-148 and the other two cases without counter injection. This result is consistent with the results obtained from the mean heat release rate profile experiments. For example, the NOx emission is below 10 PPM for injection plates A-200-69 and S-150-148 at an equivalence ratio 0.8. For the other three cases, the NOx emission is about 20 PPM at the same equivalence ratio.

Figure 12 shows the experimental results at a higher air flow rate (27.9SCFM). At this higher flow rate, the NOx concentrations are similar within the lean fuel region for all five cases. This is interpreted to mean that the fuel is mixed well with the air for all cases at higher air flow rate. More research is being performed to better understand these NOx results.

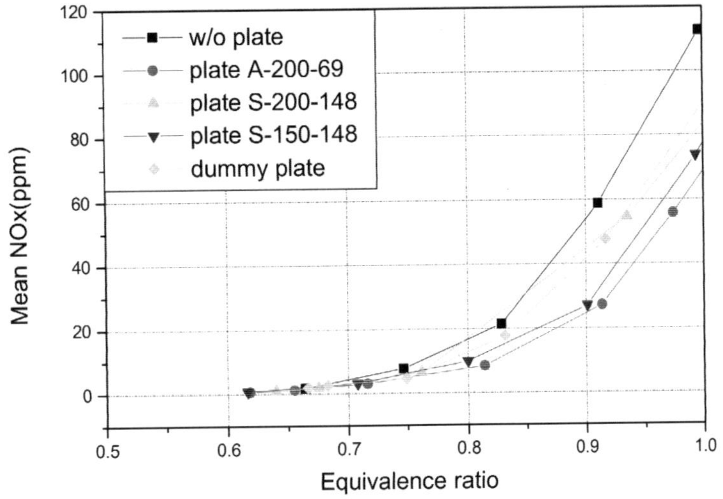

Figure 11: The mean NOx emission for five combustion cases over a range of lean air-fuel ratios at an air flow rate 11.2SCFM

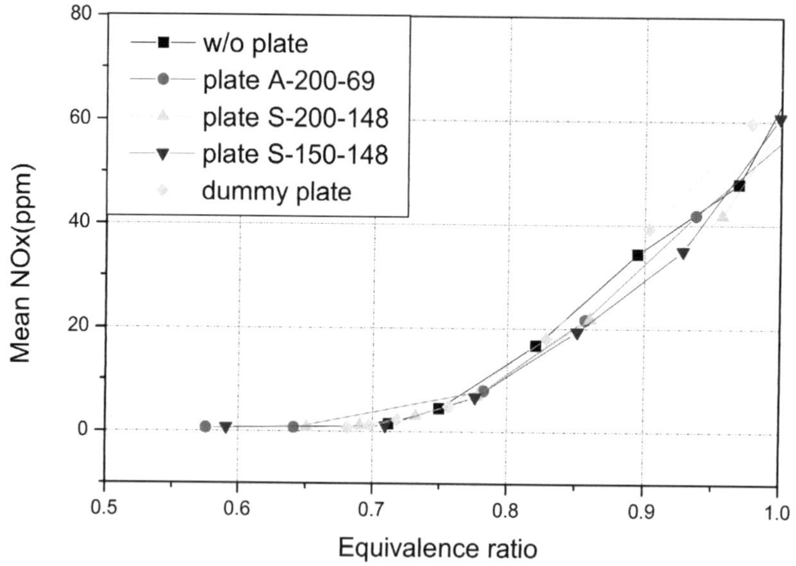

Figure 12: The mean NOx emission for five combustion cases over lean air-fuel ratios at an air flow rate 27.9 SCFM

Conclusion

A new concept multi-point micro injection premixer, has been described, fabricated and tested for use as a propane-air premixer for gas turbine applications. The design makes it possible to uniformly inject into a flow stream of gas a second gas via micro jets that are spaced apart a distance on the order of one millimeter. The micro jets can be injected in a co-flow or counterflow orientation with respect to the main gas stream.

A number of tests were performed to quantify the performance of the premixer concept. Importantly, it was found that the velocity (and mass flow) for every jet hole in the premixer was very similar, making it possible to assert that the fuel was injected out uniformly through all the jet holes regardless where a jet hole locates.

A variety of combustion experiments were also performed to evaluate the performance of the premixer concept. Among the tests performed were the equivalence ratio at LBO, the mean CH signal, and NOx as a function of air flow rate, equivalence ratio, and premixer design. In general, the premixers seemed to indicate excellent mixing is achieved, even when compared to cases involving air-fuel mixtures that were thoroughly mixed far upstream of the dump plane. In summary, the initial test results look promising, but much more research needs to be completed to better quantify the performance of the micro jet premixer.

Acknowldgments

The authors would like to thank the Louisiana Educational Quality Support Fund (LSQSF): Industrial Ties Program for supporting this research program.

References

1. C. Harris, K. Kelly, T. Wang, A. McCandless, Shahryar Motakef, "Fabrication, Modeling and Testing of Micro Cross Flow Heat Exchangers", *Journal of Microelectromechanical Systems*, Volume 11 (2002), Number 6, pp. 726-735.

2. K.W. Kelly, A.McCandless, C. Marques, R. Turner, P. Luke, S. Motakef and M. Overolt, Industrial applications of thermal devices with meso-scale features, ASME Heat Transfer Conference, San Francisco, CA, July 19, 2005.

3. J. Zhang, "Micro IOnjectionFuel-Air Premixe, Ph. D. Thesis, Louisiana State University, January 2007.

4. S. P. Nandula, R.W. Pitz, R.S. Barlow and G.J. Fiechtner, " Rayleigh/Raman/LIF measurements in a turbulent lean premixed combustor", Advanced turbine systems annual program review", Morgantown, West Virginia, Oct., 1995.

5. R. M. Flores, V.G. Mcdonell and G.S. Samuelsen, " Impact of ethane and propane variation in natural gas on the performance of a model gas turbine combustor", Journal of engineering for gas turbines and power, vol. 125, pp. 701-708, July 2003.

6. International Mezzo Technologies website: www.mezzotech.com

7. Chad Harris, "Design, fabrication, and testing of cross flow micro heat exchanger", Ph.D dissertation, Mechanical Engineering, Louisiana State University, 2001

8. R. C. Steele, A. C. Jarrett, P. C. Malte, J. H. TOnouchi, and D. G. Nicol, " Variable affecting NOx formation in lean premixed combustion", Transactions of ASME, Vol. 119, pp. 102-107, January 1997

9. V. J. Lyons, "Fuel/ air nonuniformity- effect on nitric oxide Emissions", AIAA Journal, Vol. 20, pp. 660-665, 1982

Microstructured Mixing Devices: An Efficient Tool for the Determination of Chemical Kinetic Data?

K. Hecht[1), M. Kraut[2) and A. Kölbl[2)

1.) Department of Chemical Engineering, University of Utah, 50 South Central Campus Drive, Room 3290 MEB, Salt Lake City, UT 84112, http://www.che.utah.edu

2.) Institute for Micro Process Engineering (IMVT), Forschungszentrum Karlsruhe GmbH, Hermann-von-Helmholtz Platz 1, 76344 Eggenstein-Leopoldshafen, Germany, http://www.fzk.de/imvt-en

1 Introduction

Sufficiently fast mixing is prerequisite for the determination of intrinsic kinetic data. If mixing is not sufficiently fast, data that are governed or influenced by mass transport phenomena are obtained [1]. In order to judge the influence of the mixing process on the obtained data of kinetic studies, the mixing devices need to be examined. For macroscopic dimensioned mixers, various characterization methods are available. Due to the small dimensions of microstructured devices, the choice of characterization methods is restricted or often more difficult to achieve [2].

Besides dye stuff methods like acid base reactions with a color indicator, competing chemical reactions have been applied for the characterization of microstructured mixing devices. The most popular reaction scheme is the iodide iodate reaction scheme (aka "Villermaux method") [3-7]. The competing azo coupling of 1-naphthol and 2-naphthol with diazotized sulfanilic acid was applied to characterize microstructured mixers [8,9]. Both methods are parallel-competing reactions.

We have investigated the suitability of the consecutive-competitive bromination of 1,3,5-trimethoxybenzene (TMB) for the characterization of microstructured devices [10]. The reaction scheme is depicted in Figure 1.

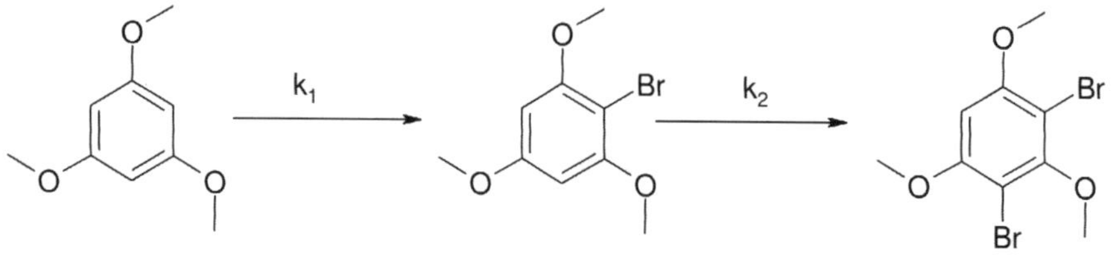

Figure 1: The Mixing Sensitive Bromination of 1,3,5-Trimethoxybenzene

Bourne and Kozicki showed in 1977, utilizing stirred laboratory flasks, that the consecutive-competitive bromination of 1,3,5-trimethoxybenzene (TMB) to 1-bromo-2,4,6-trimethoxybenzene (1BR) and 1,3-dibromo-2,4,6-trimethoxybenzene (2BR) (Figure 1) yields mixing sensitive product distributions [11]. They reacted equal volumes of 0.05M TMB in methanol with 0.05 M bromine in methanol. The better the mixing the more 1BR and the less 2BR are found in the mixing product. In case of perfect mixing, the product distribution would only be governed by the kinetics of the first and the second bromination step.

Determination of the kinetic parameters free of mass transport phenomena requires sufficiently fast mixing. Commonly used devices for the examination of fast chemical reactions are "stopped flow apparatuses" [12].

Depending on the stirring speed, Bourne and Kozicki observed product ratios of 1BR/2BR between 3 and 5.6. Furthermore Bourne and Kozicki cite mixing in a stopped-flow apparatus, where the same bromine and TMB initial concentrations were used and a product ratio of 1BR/2BR of 9.7 was achieved [13].

2 Experimental

We have investigated the bromination reaction at the same initial concentrations as Bourne and Kocziki [11] in a static cyclone type mixer. Due to the corrosive character of bromine even at very low concentrations the mixer device was manufactured in PTFE. Details of the cyclone type mixer can be seen in Figure 2 a-e. The height of each feed loop is 103 µm; the height of the exit loop is 145 µm. A detail of the feed loop feeding into the swirl chamber is shown in Figure 2c. The swirl chamber is 0.5 mm in diameter and 5 mm long. The feed loops fed into the swirl chamber in such a way that a three-dimensional rotational flow is created in the swirl chamber.

Figure 2a-e:
a: Cyclone mixer schematic
b: Cyclone mixer and housing; top plate is the exit, bottom plate contains the feeds
c: Optical micrograph of the edge of feed loop to the swirl chamber
d: Assembled microstructured mixing device
e: swirl chamber and feed loop schematic

The resulting reaction mixtures were analyzed via HPLC. A Perkin Elmer Series 200 HPLC equipped with an autosampler, a vacuum degasser, a pump, a Peltier column oven, a diode array detector, and an Agilent Zorbax SB-C18 (4.6 x 150mm, 3.5 μM) column was used to separate mixtures of TMB, 1BR, 2BR. UV absorption at a wavelength of 207 nm (20 nm bandwidth, 360 nm reference) was monitored. Water (ultrafiltered) and methanol (Merck p.A.) in a 1:1 volumetric ratio was used as a mobile phase. The oven was maintained at 20 °C. A 5 μL injection volume and a flow rate of 1 mL/min in the mobile phase were used.

The experimentally obtained product ratios 1BR/2BR and 2BR/1BR respectively over the total mass flow are depicted in Figure 3a and b.

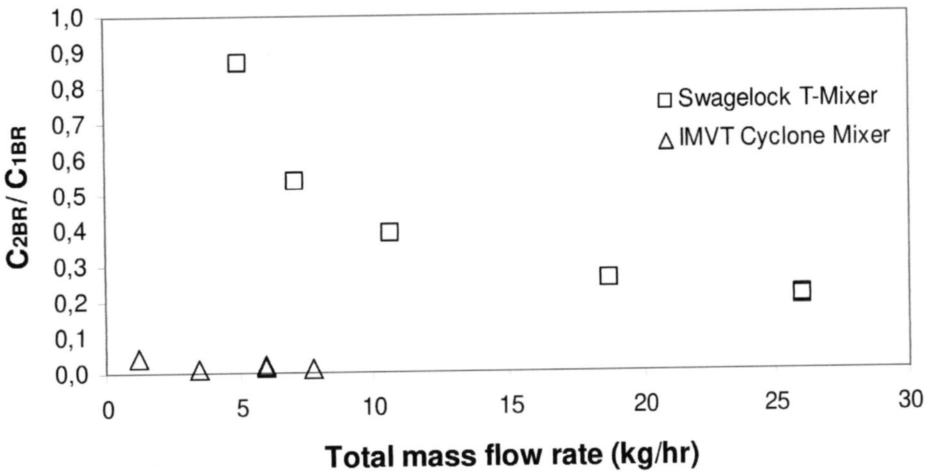

Figure 3a: Experimentally obtained 2BR/1BR ratios.

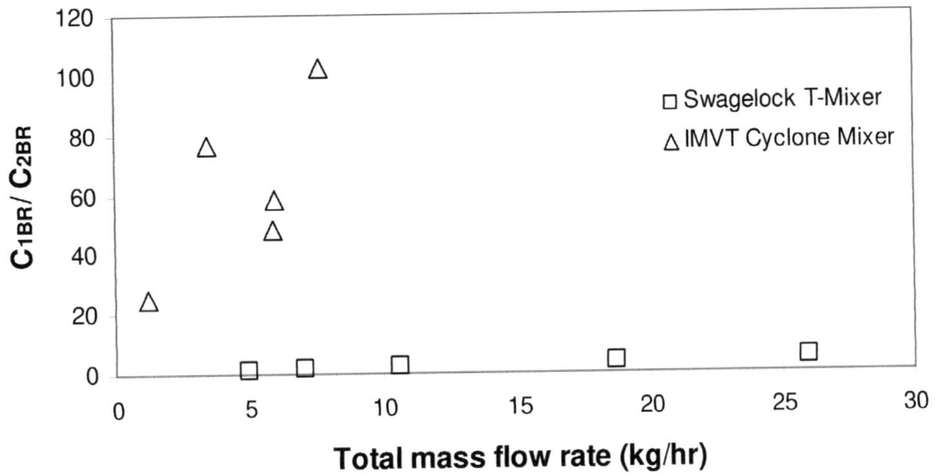

Figure 3b: Experimentally obtained 1BR/2BR ratios.

For comparison purposes the experimental results of a swagelock t-piece (nylon NY-6MO-3, inner diameter = 4.8 mm) are additionally shown in Figure 3 a and b. The obtained product ratios show better mixing results with increased mass flow indicating higher energy dissipation.

Due to the very small amount of 2BR formed in the cyclone mixer, the C1BR/C2BR ratio demonstrated large variability. This variability indicates a limit of this method. A better flow control system might provide better results. Imperfectly matched flow rates contributed to the variability of the mole balances.

3 Conclusions

Bourne and Kocziki relate the product ratio of 9.7 obtained at stopped flow experiments [13] with a ratio of $k_1/k_2 = 27$ [11]. We have derived a value of 17. The extensive derivation is given in the appendix. According to equation (16), the product ratio of 1BR/2BR = 100 obtained with the microstructured cyclone type mixer corresponds with a minimum relation of the kinetic constants k_1/k_2 of 369 in case the obtained experimental data are not affected by mass transport phenomena. Most likely the intrinsic kinetic ratio is much higher, since an estimation based on the Hammett equation is in the order of 1000 [11].

Due to the great increase in the experimentally obtained product ratio compared to the use of a stopped flow apparatus [13], we propose the use of a microstructured cyclone type mixer for the evaluation of chemical kinetic data.

4 Literature

[1]
A. Roessler and P. Rys:
"Selektivität mischungsmaskierter Reaktionen; Wenn die Rührgeschwindigkeit die Produktverteilung bestimmt"
Chemie in unserer Zeit 35(2001)314-323.

[2]
V. Hessel, H. Löwe, A. Müller und Gunther Kolb:
"Chemical Micro Process Engineering-Processing and Plants"
Wiley-VCH, Weinheim, 2005.

[3]
P. Guichardon and L. Falk:
"Characterisation of micromixing efficiency by the iodide-iodate reaction system. Part I: experimental procedure"
Chemical Engineering Science 55(2000)4233-4243.

[4]
P. Guichardon, L. Falk and J. Villermaux:
"Characterisation of micromixing efficiency by the iodide-iodate reaction system. Part I: kinetic study"
Chemical Engineering Science 55(2000)4245-4253.

[5]
M.-C. Fournier, L. Falk, J. Villermaux:
"A new parallel competing reaction system for assessing micromixing efficiency – experimental approach"
Chemical Engineering Science 51(1996)5053-5064.

[6]
M.-C. Fournier, L. Falk, J. Villermaux;
„A new parallel competing reaction system for assessing micromixing efficiency – determination of micromixing by a simple mixing model“
Chemical Engineering Science 51(1996)5187-5192.

[7]
J. Villermaux, L. Falk and M.-C. Fournier:
"Potential Use of a New Parallel Reaction System to Characerize Micromixing in Stirred Reactors“
AICHE Symposium Series 90(1994)50-54.

[8]
S. Ehlers:
"Einsatzpotentiale von Mikrosstrukturmischern“
PhD thesis, University of Cologne, Germany 1998.

[9]
S. Ehlers, K. Elgeti, T. Menzel und G. Wießmeier:
"Mixing in the offstream of microchannel systems“
Chemical Engineering and Processing 39(2000)291-298.

[10]
"Characterization of mixing processes through the consecutive-competitive bromination of 1,3,5-trimethoxybenzene"
M.S. thesis, University of Utah, USA, 2007.

[11]
J.R. Bourne and F. Kozicki:
"Mixing Effects during the Bromination of 1,3,5-Trimethoxybenzene"
Chemical Engineering Science 32(1977)1538-1539.

[12]
P.W. Atkins and J. de Paula:
"Physical Chemistry"
Oxford University Press, Oxford, New York, 2006.

[13]
P. Eugster:
"Reaktions- und diffusionsbestimmte Selektivität bei der Bromierung von Aromaten"
PhD thesis, ETH Zürich, Switzerland, 1973.

5 Appendix

$$A + B \xrightarrow{k_1} R$$
$$R + B \xrightarrow{k_2} S$$

Figure 4: Consecutive-competitive reaction scheme

The consecutive-competitive reaction scheme representing the reaction scheme depicted in Figure 1, is shown in Figure 4. Assuming elementary reaction rates, the rate of each reaction can be expressed:

$$r_A = \frac{dC_A}{dt} = -k_1 C_A C_B \tag{1}$$

$$r_S = \frac{dC_S}{dt} = k_2 C_R C_B \tag{2}$$

$$r_R = \frac{dC_R}{dt} = k_1 C_A C_B - k_2 C_R C_B \tag{3}$$

Dividing Equation 1 and Equation 2 by Equation 3 yields Equation 4 and Equation 5:

$$\frac{dC_S}{dC_A} = \frac{k_2 C_R}{-k_1 C_A} \tag{4}$$

$$\frac{dC_R}{dC_A} = -1 + \frac{k_2 C_R}{k_1 C_A} \tag{5}$$

Equation 6 can be written due to stoichiometry.

$$C_A + C_R + C_S = C_{A0} \tag{6}$$

Equation 6 can be substituted in Equation 4 to yield Equation 7.

$$\frac{dC_S}{dC_A} + \left(\frac{-k_2}{k_1 C_A}\right) \cdot C_S = \frac{k_2}{k_1} \cdot \left(1 - \frac{C_{A0}}{C_A}\right) \tag{7}$$

Equation 5 can be rearranged to give Equation 8.

$$\frac{dC_R}{dC_A} + \left(\frac{-k_2}{k_1 C_A}\right) \cdot C_R = -1 \tag{8}$$

Equation 7 and Equation 8 are first-order linear differential equations. A first-order linear differential equation and its solution are represented in Equation 9 and Equation 10, where K is a constant of integration.

$$\frac{dy}{dx} + P(x) \cdot y = Q(x) \tag{9}$$

$$y \cdot e^{\int P(x)dx} = \int Q(x) \cdot e^{\int P(x)dx} dx + K \tag{10}$$

Solving Equation 7 and Equation 8 according to Equation 9 and Equation 10 yields Equation 11 and Equation 12.

$$C_S \cdot e^{\left(\frac{-k_2}{k_1} \cdot \ln(C_A)\right)} = \frac{-k_2}{(k_2 - k_1)} \cdot C_A^{\left(1-\frac{k_2}{k_1}\right)} + C_{A0} C_A^{\frac{-k_2}{k_1}} + K_1 \tag{11}$$

$$C_R \cdot e^{\left(\frac{-k_2}{k_1} \cdot \ln(C_A)\right)} = -\frac{C_A^{\left(1-\frac{k_2}{k_1}\right)}}{\left(1 - \frac{k_2}{k_1}\right)} + K_2 \tag{12}$$

Constants of integrations K_1 and K_2 can be solved by applying the initial boundary condition that C_S and C_R equal 0 and C_A equals C_{A0}. Expressions for C_S and C_R in terms of C_A, C_{A0}, k_1, and k_2 are obtained as shown in Equation 13 and Equation 14.

$$C_S \cdot C_A^{\frac{-k_2}{k_1}} = \frac{-k_2}{(k_2 - k_1)} \cdot C_A^{\left(1-\frac{k_2}{k_1}\right)} + C_{A0} C_A^{\frac{-k_2}{k_1}} + \frac{k_1}{(k_2 - k_1)} \cdot C_{A0}^{\left(1-\frac{k_2}{k_1}\right)} \tag{13}$$

$$C_R \cdot C_A^{\frac{-k_2}{k_1}} = -\frac{k_1}{(k_1 - k_2)} C_A^{\left(1-\frac{k_2}{k_1}\right)} + \frac{k_1}{(k_1 - k_2)} C_{A0}^{\left(1-\frac{k_2}{k_1}\right)} \tag{14}$$

Equation 13, which expresses the ratio of C_R to C_S in terms of the ratio of k_2 to k_1 and the ratio of C_{A0} to C_A, is obtained by dividing Equation 14 by Equation 13.

$$\frac{C_R}{C_S} = \frac{\left(\frac{C_{A0}}{C_A}\right)^{\left(1-\frac{k_2}{k_1}\right)} - 1}{\frac{k_2}{k_1} + \left(1 - \frac{k_2}{k_1}\right) \cdot \frac{C_{A0}}{C_A} - \left(\frac{C_{A0}}{C_A}\right)^{\left(1-\frac{k_2}{k_1}\right)}} \tag{13}$$

In a stoichiometric mixture, B is consumed entirely in the reactions, so Equation 14 can be written, and Equation 15 can be obtained from Equation 6 and Equation 14.

$$C_{B0} = C_{A0} = C_R + 2C_S \tag{14}$$

$$\frac{C_{A0}}{C_A} = \frac{C_R}{C_S} + 2 \tag{15}$$

If Equation 15 is substituted into Equation 13, Equation 16 is obtained. If the ratio of k_2 to k_1 and the ratio of C_R to C_S are each considered to be a variable, Equation 16 contains only 2 variables, and the ratio of k_2 to k_1 can be obtained when the product ratio, C_R to C_S, is known for a kinetically limited system (where mass transport and mixing do not affect the reaction).

$$\frac{C_R}{C_S} = \frac{\left(\frac{C_R}{C_S} + 2\right)^{\left(1-\frac{k_2}{k_1}\right)} - 1}{\frac{k_2}{k_1} + \left(1 - \frac{k_2}{k_1}\right) \cdot \left(\frac{C_R}{C_S} + 2\right) - \left(\frac{C_R}{C_S} + 2\right)^{\left(1-\frac{k_2}{k_1}\right)}} \tag{16}$$

Theoretical and Experimental Studies of Mixing Enhancement in Micromixers

John T. Adeosun and Adeniyi Lawal. Dept. of Chemical, Biomedical, and Materials Engineering, New Jersey Center for MicroChemical Systems, Stevens Institute of Technology, Castle Point on Hudson, Hoboken, NJ 07030

Mixing in microchannel mixers/reactors, most especially for liquid phase reactions, is a relevant technical issue critical to the development and application of integrated microchemical processing systems. The objective of this work is to investigate mixing enhancement in microchannel mixers, through a theoretical as well as an experimental mixing study of currently utilized and three proposed micromixing configurations. The proposed micromixers are passive mixers uniquely designed to effect mixing using static or passive mixing structures by creating folding as well as local and global re-orientation of fluid interfaces. Using Computational Fluid Dynamics (CFD) approach and suitable mixing characterization techniques, one of the proposed mixing configurations herein referred to as multilaminated/elongational flow micromixer-4 (MEFM-4) was selected based on the set criteria of minimum pressure drop and high mixing performance. The selected MEFM-4 was then employed along with the standard T-junction micromixer (TjM) for our experimental mixing study. In order to experimentally characterize mixing in the micromixers, residence time distribution (RTD) measure in conjunction with UV-vis absorption spectroscopy detection technique was used. Using the so-called convolution-deconvolution technique based on Levenspiel's Convolution Integral Theorem, some of the challenges associated with the use of time-dependent concentration data of a dye tracer for RTD determination in micromixers were overcome. With the application of this technique and suitable programming scheme, the experimental data obtained for the T-junction micromixer and MEFM-4 were analyzed while dispersion model and other suitable models were used to fit the data. The model parameters obtained were then used to indirectly evaluate mixing quality in the micromixers at different flow conditions. The validation of our prior theoretical results and comparison of the mixing performance of the standard TjM with the proposed MEFM-4 will also be presented.

Re-dispersion Microreactor to Achieve Staged Liquid-Liquid Dispersion for Use in Polycondensation

Volker Hessel[1,2], Sonja Rothstock[1], David W. Agar[3], Prasad Jadhavrao[1,3], Patrick Löb[1], Bernd Werner[1]

[1] Institut für Mikrotechnik Mainz GmbH, Mainz, Germany
[2] Eindhoven University of Technology, Eindhoven, Netherlands
[3] University of Dortmund, Dortmund, Germany.

To ensure mass transfer in biphasic reactions a sufficient interface area is required. Within the scope of micro process engineering two concepts are available. The first one is the continuous-phase contacting and the second one is the dispersing mixing technique. The latter can be realised in a micromixer-capillary reactor combination, which can successfully used, if the emulsions are long-term stable [1, 2]. An example for such application is the transient catalyst screening for the isomerisation of allylic alcohols [3].

In some cases, an emulsifier is added to stabilise the interface. In this case often a single-point provision of energy is sufficient to form and preserve an emulsion. In many other cases, however, addition of emulsifiers is prohibited, since that may negatively affect product purification or quality. The then correspondingly faster coalescence of the droplets poses a challenge to the micromixer technology. Because dispersions of various immiscible solvents typically used for organic synthesis are stable at least for some seconds, a reactor concept has been developed avoiding such energy-intensive solutions. This so-called re-dispersion microreactor contains exchangeable serial re-dispersion units and an observation window for flow inspection. The re-dispersion units are separated by multichannels with stage-wise increasing channel length.

For one first model system investigated, the heptane / water (SDS) system, droplet size spectra were gained by a droplet analyser.

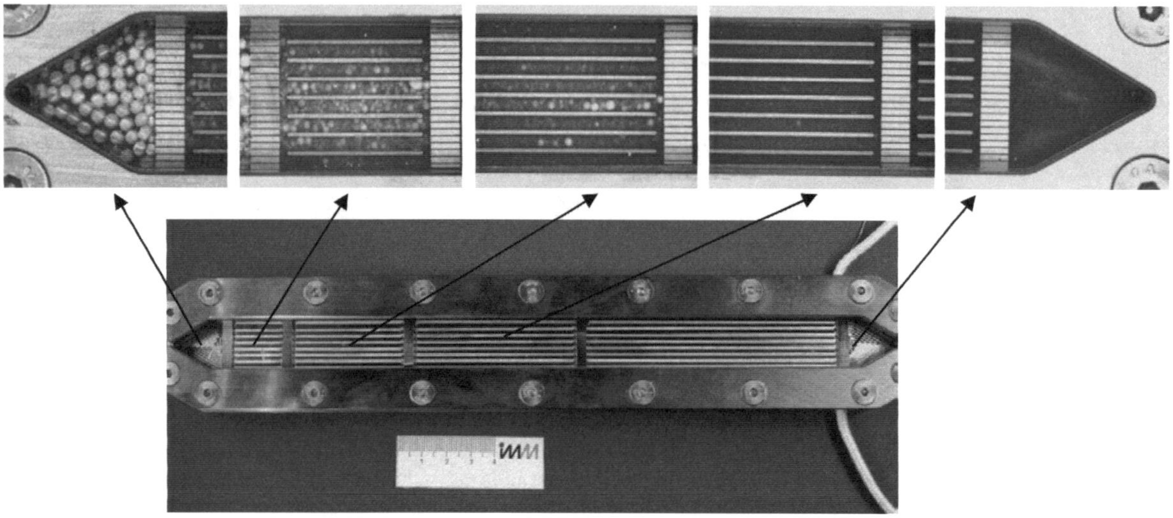

Figure 1. *Model system heptane / water (SDS), droplet refining.*

In the process flow, the solvents are initially roughly pre-mixed by a standard caterpillar micromixer, resulting in large droplets when entering the inlet chamber of the re-dispersion reactor. The crude dispersion is then reduced to much smaller size already by pass through the re-dispersion units, finally forming in the outlet segment a fine emulsion.
A decrease in average droplet size with increasing flow rate is found. The same tendency can be observed for symmetric (1:1) flow ratios as opposed to asymmetric (1:5) ones. Two different kinds of redispersion units are used. The standard microchannel redispersion unit contains 20 parallel microchannels (width 0.1 mm, depth 0.1 mm), the second one is made of metal foam with an average pore diameter of 0.4 mm and a specific surface area of 5400 m^2/m^3. For small flow rates (under 500 mL/h) the standard microchannel redispersion units give smaller average droplet diameter in comparison with the metal foam. Admittedly the metal foam redispersion unit generates a lower pressure drop, consequently less specific energy is needed to create the same average droplet diameter compared to the standard microchannel.

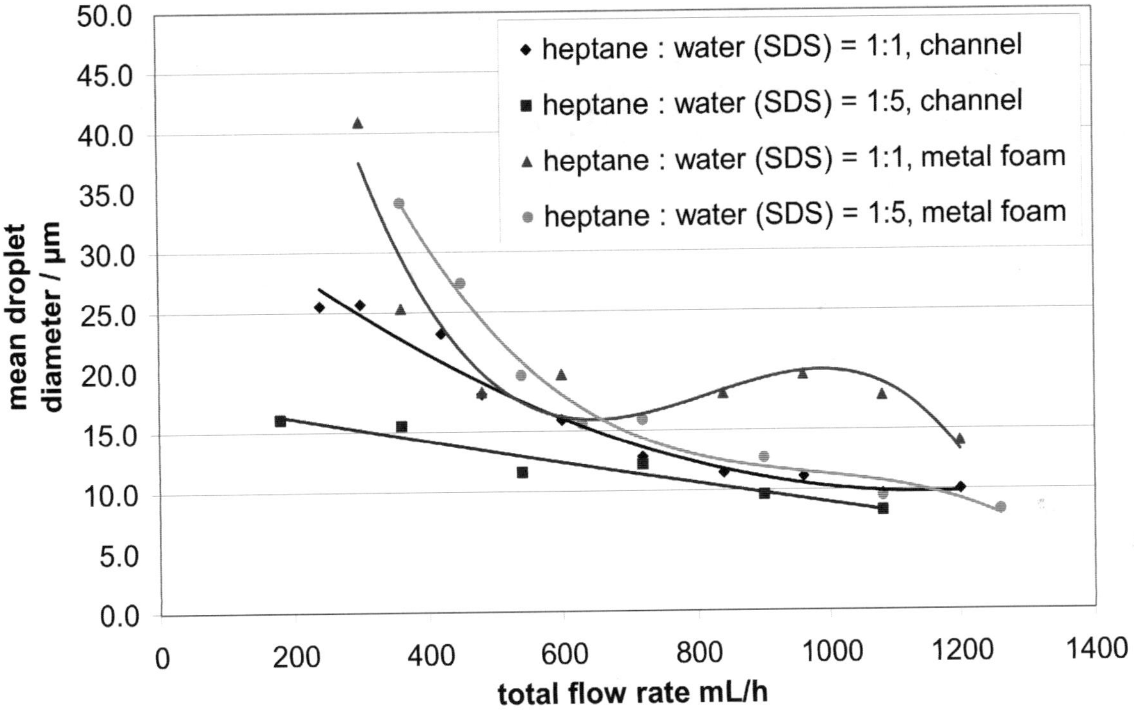

Figure 2. *Comparison of standard channel and metal foam redispersion units using the model system heptane / water (SDS)*

To judge on the performance of this special microreactor a comparative study using standard interdigital micromixers with heptane / water (SDS) was accomplished. These devices show a decrease of the average droplet size with energy, respectively flow rate, and flow compression through a nozzle-like geometric confinement ('slit'). For a heptane / aqueous phase ratio of 1/5 and small flow rates the redispersion reactor is more energy efficient.

A subsequent real-case solvent system of toluene: dioxane: water without addition of an emulsifier is investigated in the new developed re-dispersion reactor. This system constitutes the solvent mixture for a polycondensation reaction to synthesize OLED materials. Here, high demands are posed on conversion and product purity. When dispersing the immiscible phases at the re-dispersion units, various flow regimes varying from finely dispersed phase to drop, slug and deformed interface flow are observed. Specific interfacial area and its energy requirement were calculated at different flow rates for the real case system, giving an enhanced interfacial area and lower power requirement by a factor of 10 than the conventional equipments used for contacting two immiscible phases in chemical industry.

The metal foam re-dispersion units appeared to be advantageous regarding the flow distribution, interfacial area and the specific energy requirement. The specific interfacial area at different flow rates was calculated by taking photographs from a high speed camera.

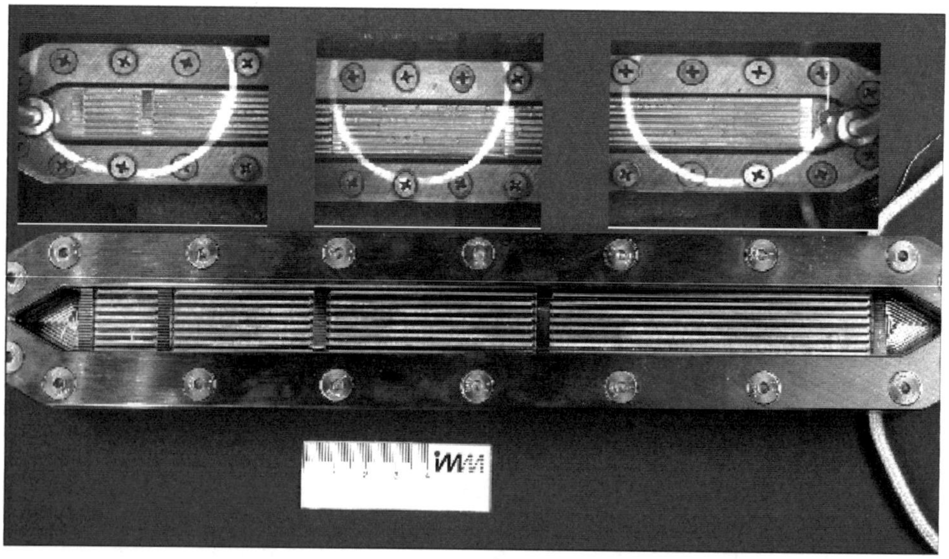

Figure 3. *Analysed photographes, solvent system: toluene : dioxane : water*

The organic phase is dispersed in the aqueous phase as the latter wets the steel channels of the re-dispersion reactor. The photographs were then analyzed by using Image Tool. These 2D photographs were then converted into 3D in order to calculate its surface area. Three random photographs were chosen and each photograph was analyzed three times, the mean deviation in each photograph analysis was less than 5%. The average surface area of three analyzed photographs was used to calculate the specific interfacial area. The specific surface area at a particular flow rate was calculated by adding the surface areas of the dispersed organic phase at the given snapshots in all segments of the re-dispersion reactor to the total reaction volume available for dispersion. The following diagram shows the variation of specific interfacial area at different average flow velocities for metal foam and channel re-dispersion units at 20°C and 1 bar.

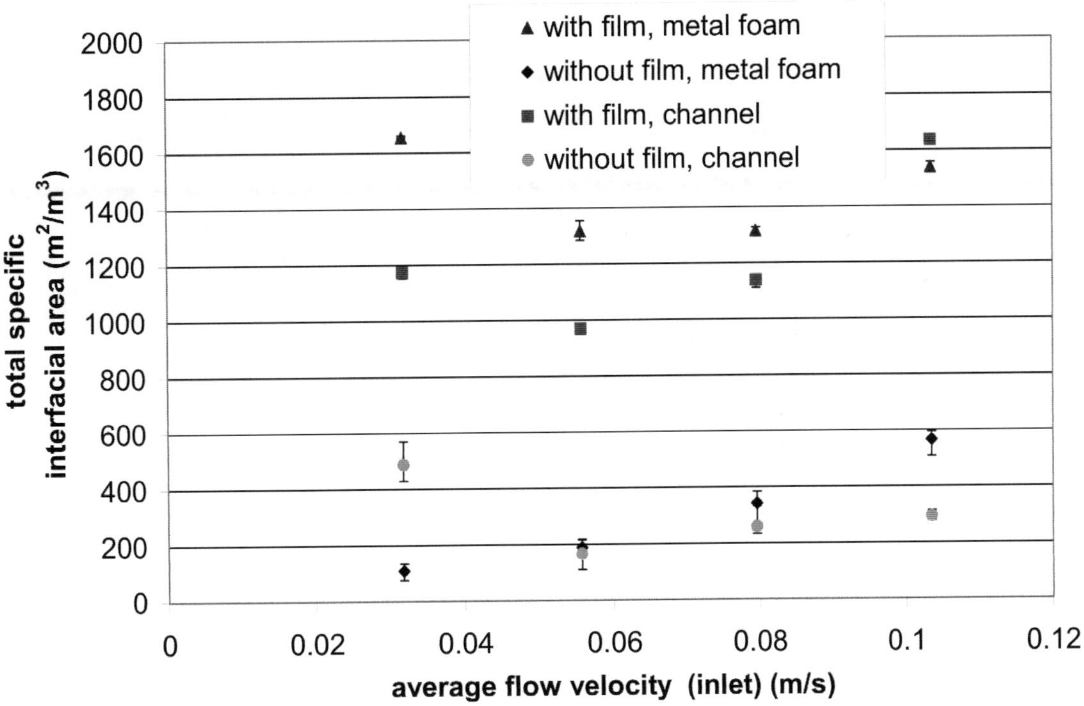

Figure 4. *Comparison of the total specific interfacial areas for the standard channel and metal foam redispersion units with and without film*

The specific interfacial area is calculated by two different ways, first considering a presence of an aqueous film through which the organic phase slug flows and second considering without film [4]. It is seen from the figure that as the average flow velocity at the re-dispersion reactor inlet increases the interfacial area goes through a minimum which can be attributed to the phenomenon of coalescence. However at lower flow velocity drops are created at the re-dispersion units and they migrate into the non-dispersed channels leading to the increase in channel filling and specific interfacial area, as at lower flow velocities not all the channels in the re-dispersion reactor are filled. Moreover the interfacial area increases with average flow velocity after the minimum due to better flow distribution and presence of more organic content.

The range of interfacial area values obtained is 900-1650 m^2/m^3 calculated with film which is around 10 times higher than the conventional equipment used for contacting immiscible fluids. Even the interfacial area values obtained without film are still greater than for some mixing devices used in chemical industry. The power requirement for producing the given specific area for the foam re-dispersion units is between 4-20 kJ/m^3, which is by factor 10-100 less than required by conventional mixing equipments. [5].

References:

[1] Löb, P., Hessel, V., Men, Y., Pennemann, H. *Chem. Eng. Sci.* **61** (2004) 2959-2967.

[2] Pennemann, H., Hardt, S., Hessel, V., Löb, P., Weise, F. *Chem. Eng. Tech.* **28**, 4 (2005) 501-508.

[3] de Bellefon, Tanchoux, N., Caravieilhes, S., Grenouillet, P., Hessel, V. *Angew. Chem.* **112**, 19 (2000) 3584-3587.

[4] Kashid M.N, Agar D, *J. Chem. Eng.,* in press, available online 3 January 2007

[5] Dehkordi, A. M., *Ind. Eng. Chem. Res.* **40** (2001), 681 – 688.

Stable Disrupted Flow in Microchannels for Multiphase Reactions

Anna Lee Tonkovich[1], Maddalena Fanelli[1], Dongming Qiu[1], Rick Stevenson[2], Amanda Glass[2], Jan Lerou[2], and Laura Silva[3]. (1) Technology Development, Velocys, Inc., 7950 Corporate Boulevard, Plain City, OH 43064, (2) Experimental Operations, Velocys, Inc., 7950 Corporate Boulevard, Plain City, OH 43064, (3) Business Development, Velocys, Inc., 7950 Corporate Boulevard, Plain City, OH 43064

Multiphase reactions have been notoriously challenging to perform in microchannel reactors. Recent advancements in the literature have suggested the use of Taylor flow as one method to enhance contact and reaction between phases. An alternate approach to efficient multiphase mixing has been developed to disrupt flow and enhance mixing in otherwise laminar flow microchannels. Stable disrupted flow is created by the integration of wall depressions or features which break the laminar flow regime within a microchannel. The result is turbulent-like flow in otherwise low Reynolds number systems. This approach extends beyond the original work for liquid-liquid mixing described by Svasek et al in 1996, Johnson et al in 2002, and Strook et al in 2002 who added grooved wall features to enhance mixing in microfluidic systems with very low Reynolds numbers. The microchannel design has evolved to address additional challenges of multiphase flow. Comparisons of conventional microchannel performance with those employing stable disrupted multiphase flow will be made using both numerical simulations and experiments.

Johnson, Timothy J., Ross, David and Locascio, Laurie E., "Rapid Microfluidic Mixing," Analytical Chemistry, Vol. 74, No. 1, January 1 (2002).

Stroock, Abraham D., Dertinger, Stephan K.W., Ajdari, Armand, Mezic, Igor, Stone, Howard A., and Whitesides, George M., "Chaotic Mixer for Microchannels," Science Vol. 295, 25 January (2002).

Svasek, Peter, Jobst, Gerhard, Urban, Gerald and Svasek, Edda, "Dry Film Resist Based Fluid Handling Components for μTAS," Analytical Methods & Instrumentation: Special Issue, 78-80, (1996).

Fabrication of Discotic Colloids through Electrospray

Dawei Luo[1], Manuel Marquez[2], and Zhengdong Cheng[1]. (1) Chemical Engineering, Texas A&M University, College Station, TX 77840, (2) Arizona State University; National Institute of Standards and Technology; Los Alamos National Laboratory

In this study, we pressent a novel mass production method of discotic wax particles using electrospray. The 1-eicosene wax was melt and added with an electrical conductive additive to increase its electric conductivity. Liquid wax was then flow through a metallic nozzle with applied high voltage, and broke into fine droplets, which were collected using SDS aqueous solution. After storage at 4 C for several weeks, wax particles experienced a shape transition into microdisks.

Enhancement of non-Newtonian transport behaviour of emulsions in micro tubes and channels at high content of dispersed (incoherent) phase

Tobias Sprogies [1)], G. Alexander Gross [1)], Daniell Malsch [1,2)], Thomas Henkel [2)] and J. Michael Köhler [1)]

1) Technical University of Ilmenau, Institute of Micro and Nanotechnologies/Institute for Physics

2) Institute of Physical High Technology Jena

correspondence:
Prof. Dr. M.Köhler, TU Ilmenau, Inst. f. Micro and Nanotechnologies, Inst. f. Physics, Weimarer Str. 32, D-98693 Ilmenau, Germany, phone 49-3677-693655, michael.koehler@tu-ilmenau.de,

Abstract

The well-known non-Newtonian fluid behaviour of emulsions is caused by shear-forced deformation of droplets (incoherent phase), by induction of phase internal convection and by shear-forced induced lateral movement of droplets. Here, the flow behaviour of oil-in-water emulsions in micro channels and capillaries at low water volume fraction was studied. Silicone oil was applied as non-aqueous phase, sodium dodecyl sulphate (SDS) was used as surfactant. The transport behaviour was characterized by the flow rate-depending pressure drop and by the determination of local trajectories inside micro channels using micro particle image velocimetry (µ-PIV). The emulsions were generated immediately before the measurements by flow-through emulsification using a static chip micro mixer. It was found, that the character of the transport behaviour is changed considerably, if the volume fraction of the aqueous phase is shifted below about 30%. At lower volume fraction of water, the fluid resistance decrease much stronger with increasing flow rate than at higher water content. The µ-PIV investigations show, that this effect is due to a transition from a more laminar to a strong plug-like flow regime. The plug flow character is marked by a steep velocity gradient in the liquid near the channel walls. So, the transport behaviour can be described by a significant slip-effect. In addition, the direct mechanical contact between the single oil droplets contributes to the change in the response behaviour of the emulsion against the shear stress. The investigations show that the flow of emulsions of low volume fraction of coherent phase through micro channels and tubes is strongly determined by geometrical conditions of spatial self-organization of droplets of an emulgated liquid. Beside the volume ratio itself, the viscosity of both phases, the size distribution, the flow rate and the channel width are additionally effecting the transport behaviour.

Key Words

micro reactor, emulsion, static micro mixer, slip effect, slug flow, non-Newtonian flow, self-organization, Magnus effect

1. Introduction

Heterogeneous systems like emulsions and suspensions show frequently a non-Newtonian flow behaviour [1,2]. The effect of shear forces on the viscosity causes changes in the pressure drop during the liquid transport. The flow resistance of processed liquids is of central importance for microfluidic systems and microreaction technology. Liquids of higher viscosity and multiphase systems can cause strong pressure drops if they are moved through small micro channels [3]. Designs of microreactors must be carried out with respect to the rheological properties of the reactant liquids and the changes of viscosity during the processes inside the reactor or the micro fluidic system.

The pressure drops can be estimated easily by the law of Hagen-Poiseuille for tubes with circle cross sections, if the viscosity of the streaming liquids are known. The pressure drop is proportional to the viscosity for regular fluidic behaviour (Newtonian liquids). But, there is a lot of applications of liquids in microsystems showing non-Newtonian flow behaviour. The resulting flow resistance of these liquids depends on the response of liquids on the local shear forces. These forces are dependent on flow rate and on the geometry of fluidic channels. In microsystems, particular high shear rates result from the small channel dimensions. So, the specific effects of transport behaviour of emulsions should be of particular importance in microfluidics and for microreactor applications. Emulsions play a crucial role in many technical processes. Extraction processes, phase transfer reactions and emulsion polymerization are some examples for the application of emulsions. Emulsions are of importance in food industry, in cosmetics, pharmacy and other fields.

It was shown earlier, that micro reactors can be applied for the generation of emulsions [4]. In particular, multilamination mixers and other types of static micro mixers are well suited for the generation of emulsions [5,6]. In most cases, one of the both liquids, namely the dispersed phase, was applied in a certain excess. So, the particles of the incoherent phase are isolated from each other by the embedding liquid. But, in many applications, high concentrations of the dispersed phase (embedded phase, incoherent phase) are needed. Therefore, here the effect of generation and transport of emulsions with low content of the coherent phase (dispersed phase) was studied.

2. Experimental

Silicone oil (viscosities of 4.9 mPas and 19.5 mPas) and an aqueous solution of sodium dodecylsulphate (SDS, 5%) were chosen as model substances in the experiment. The emulsions were either prepared in situ or before the measurements. The emulsions were prepared by use of a static micro mixer (Statmix 6, IPHT Jena [7]). Syringe pumps are applied for actuation of both solutions (fig. 1a). Flow rates between 30 and 800 µL/min were used for the generation of emulsions. The product liquid consisted of three parts (emulsion, oil, aqueous solution), which could be separated by density.

The pressure measurements were carried out by a flow-through arrangement consisting of two syringe pumps, a three-way valve, two pressure sensors and the measurement tube (fig. 1b). At the beginning, the measurement tube was filled with aqueous solution of surfactant. Then the syringe with emulsion was switched on and the valve was moved into the emulsion position. The flow rate was determined by the emulsion syringe during the measurement. The flow resistance of tubes was calculated by the measured difference of pressure at the both sensor position and the applied flow rate. In some experiments, the generation of emulsions and the measurements were carried out together. Therefore, the formed emulsion was fed directly from the mixer to the pressure measurement arrangement. In this case, variations of flow rates did not only effect the pressure drop but also the formation of emulsion itself.

The emulsions were inspected by light microscopy. The main fraction of oil droplets was in the diameter range of about 15 µm in case of silicone oil of higher viscosity. An additional shoulder in the size distribution spectrum was found in the range of 30 µm in case of the oil of lower viscosity (fig. 2). This shoulder represents the largest volume fraction of droplets in the emulsions of the oil with lower viscosity.

The velocity distribution during the transport of emulsions through microchannels was investigated by micro particle imaging velocimetry (µ-PIV). These measurements were based on a series of microscopic images of streaming liquid taken from a transparent microfabricated channel chip (glass/glass).

3. Results and discussion

3.1. Flow behaviour of preformed emulsions

The pressure drop was recorded after starting of experiments for about 2 minutes. In all cases, a shift of pressure drop was observed after the start of the emulsion transport. This shift could be interpreted as a reduction of apparent viscosity. The effect increases with increasing flow rates (fig. 3). The final apparent viscosity was achieved in a time interval less than 1 second at a flow rate of 0.2 mL/min. It takes more than 10 s at a flow rate of 3.2 mL/min. The effect of time-dependent pressure drop is obviously due to a reorganisation of droplets inside the emulsion after application of flow. It is probably caused by a rearrangement of oil spheres under the shear stress of streaming emulsion. Increasing flow rate (increasing shear stress) causes a longer time and a stronger effect of reorganization.

It was expected, that effects of reorganization would depend on the interaction between the oil droplets. The probability of such interactions increases with increasing oil fraction inside emulsion. A strong non-linear dependence of apparent viscosity with increasing oil fraction was observed for all flow rates (fig. 4). This effect can not only be explained by the higher viscosity of the oil in comparison with water but would be due to the increase in the intensity of droplet interactions. The increase of apparent viscosity is high at lower flow rates and decreases with increasing shear forces. This observation seems to reflect the importance of shear stress-induced reorganization of droplets. A particular high increase in flow rate effect on apparent viscosity was found in the volume range between 60 Vol% and 75 Vol% of oil (fig. 5). The pressure drop measurements show that emulsions with a high content of oil (around 70% and more) possess a significant different flow behaviour than emulsions of lower oil content.

The conclusions on results of pressure drop experiments were supported by the μ-PIV measurements. Strong differences between the velocity distribution inside a micro channel were found between emulsions of 30% and 70% oil at a flow rate of 30 μL/min. The μ-PIV profile for the emulsion with low oil content is marked by a continuous increase of flow rate from the walls to the central region of the channel (the large asymmetry is caused by gravity influence in connection with the experimental arrangement). Despite the asymmetry, the profile shows a certain relation to a parabolic profile like known from classical flow conditions. The high content emulsion shows a completely different velocity profile at same flow conditions in the same experimental arrangement (fig. 6a). The μ-PIV measurements describe a steep increase in liquid velocity (high velocity gradients) near the walls but a nearly constant velocity over a wide range of the whole diameter (constant velocity). This fact can only be understood by the assumption of the formation of a comparatively rigid column of oil spheres (droplets) inside the channel. The oil droplets are in direct mechanical contact. They behave like an uniform plug. All oil droplets are moved by the same velocity. The measurements at higher flow rate matches this interpretation. The velocity distribution in a 70% oil containing solution at 800 μL/min is characteristic for a moving plug, too (fig. 6b, No2). The peripheral region of velocity increase is a little enlarged in comparison with the low flow rate. This extension can be understood by reconsidering the much higher shear forces in cause of the higher flow rate. At this high flow rate, a certain plug transport effect is also be found for lower oil concentration (fig. 6b, No1). But, the velocity gradients in the near-wall region are considerably lower than in case of an emulsion with 70% oil. (The gravity-induced asymmetry of profiles in the μ-PIV experiments is considerably lowered in case of high flow rates). The PIV profiles describe a strong slip effect for the transport of emulsions of different oil content at high flow rates. Under the shear stress of liquid motion, all emulsions are divided in a central part moving nearly with an uniform velocity ("plug") and a peripheral part ("lubricant"). This effect is caused by a shear stress-induces self organization effect inside the emulsion. The oil droplets are transported from the peripheral region into the central region. In result, the peripheral liquid layer near the walls loose oil droplets, but the emulsion becomes denser in the central part of the channel. The transport of droplets from the periphery into the centric part is not restricted to higher concentrated emulsions but is due to the Magnus effect which is generally responsible for a motion of particles and droplets exposed to a non-linear velocity gradient.

For lower and medium oil content, the spontaneous subdivision of liquid into a plug-like part in the middle and a lubricant-like part at the periphery of the channel is significant only at higher flow rates. In contrast, the analogue effect can be observed even at lower flow rates if the oil content is high.

3.2. In-situ measurements in a flow-through system during generation of emulsion

The response of pressure drop on a streaming emulsions becomes more complex if the emulsions are generated in situ. In most cases, a strong non-monotonous pressure difference signal was observed after on-switching of the emulsion formation. The strong increase in pressure signal at the beginning is followed by a reduction of apparent viscosity. Obviously, the formation as well as the transport of emulsion need a certain time to reach the steady state. The finally achieved pressure drop in this steady state depends on flow rate and on the oil content of emulsions or the initial flow rates of aqueous solution and oil, respectively (fig. 7). The decrease of pressure drop dependence on flow rate above 1 mL/min (fig. 8a) and the increase above 0.5 mL/min (fig. 8b) seems to be due to a change in mechanism of emulsion formation. The assumption of a change in mechanism is supported by the relaxation rates of apparent viscosity after the initial peak. The relaxation rates increases quickly with flow rate at lower shear forces, but increase slower above 1 mL/min (fig. 8).

4. Conclusions

The well-known non-Newtonian fluid behaviour of emulsions is caused by shear-forced deformation of droplets (dispersed phase), by induction of phase internal convection and by shear-forced induced lateral movement of droplets.

It was found, that the character of the transport behaviour is changed considerably, if the volume fraction of the aqueous phase is shifted below about 30%. At lower volume fraction of water, the fluid resistance decrease much stronger with increasing flow rate than at higher water content. The transport behaviour can be described by a significant slip-effect. In addition, the direct mechanical contact between the single oil droplets contributes to the change in the response behaviour of the emulsion against the shear stress. This effect becomes particular important, if the critical volume ratio for dense stacked micro sphere is achieved in case of a homogenous droplet size. The µ-PIV investigations showed, that this effect is due to a transition from a more laminar to a strong plug-like flow regime. The plug flow character is marked by a steep velocity gradient in the liquid near the channel walls The exact value for critical volume ratio depends on the variation of droplet sizes inside emulsion and on the flow-induced droplet deformation. The investigations show that the flow of emulsions of low volume fraction of coherent phase through micro channels and tubes is strongly determined by geometrical conditions of spatial self-organization of droplets of embedded liquid. Beside the volume ratio itself, the viscosity of both phases, the size distribution, the flow rate and the channel width are additionally effecting the transport behaviour.

Acknowledgement

We thank T. Frank (LTF Ilmenau) for cooperation and discussion, F. Möller and S. Schneider (TU Ilmenau) for support and technical assistance. Financial support by the German Environmental Foundation (DBU) is gratefully acknowledged.

References

[1] F.W.A.M. Schreuder, H.N. Stein: Rheologica Acta 26 (1987), 45
[2] J. Mewis, A.J.B. Spaull: Adv. Coll. Interf. Sci. 6 (1976), 173
[3] M.K. Jensen, Y. Peles, Th. Borca-Tascciuc, S.G. Kandlikar: Multiphase Flow, Evaporation and Condensation at the Microscale, in: N. Kockmann (ed.): Micro Proecess Engineering (Weinheim 2006), 115
[4] N. Schwesinger: privat communication
[5] W. Ehrfeld, V. Hessel, H. Löwe: Microreactors (Weinheim 2000), p 67
[6] I. Kobayashi, M. Nakajima: Generation an Multiphase flow of emulsions in microchannels, in: N. Kockmann (ed.): Micro Proecess Engineering (Weinheim 2006), 149
[7] T. Kirner, J. Albert, M. Günther, G. Mayer, K. Reinhäckel, J.M. Köhler: Chem. Engn. J. 101 (2004), 65

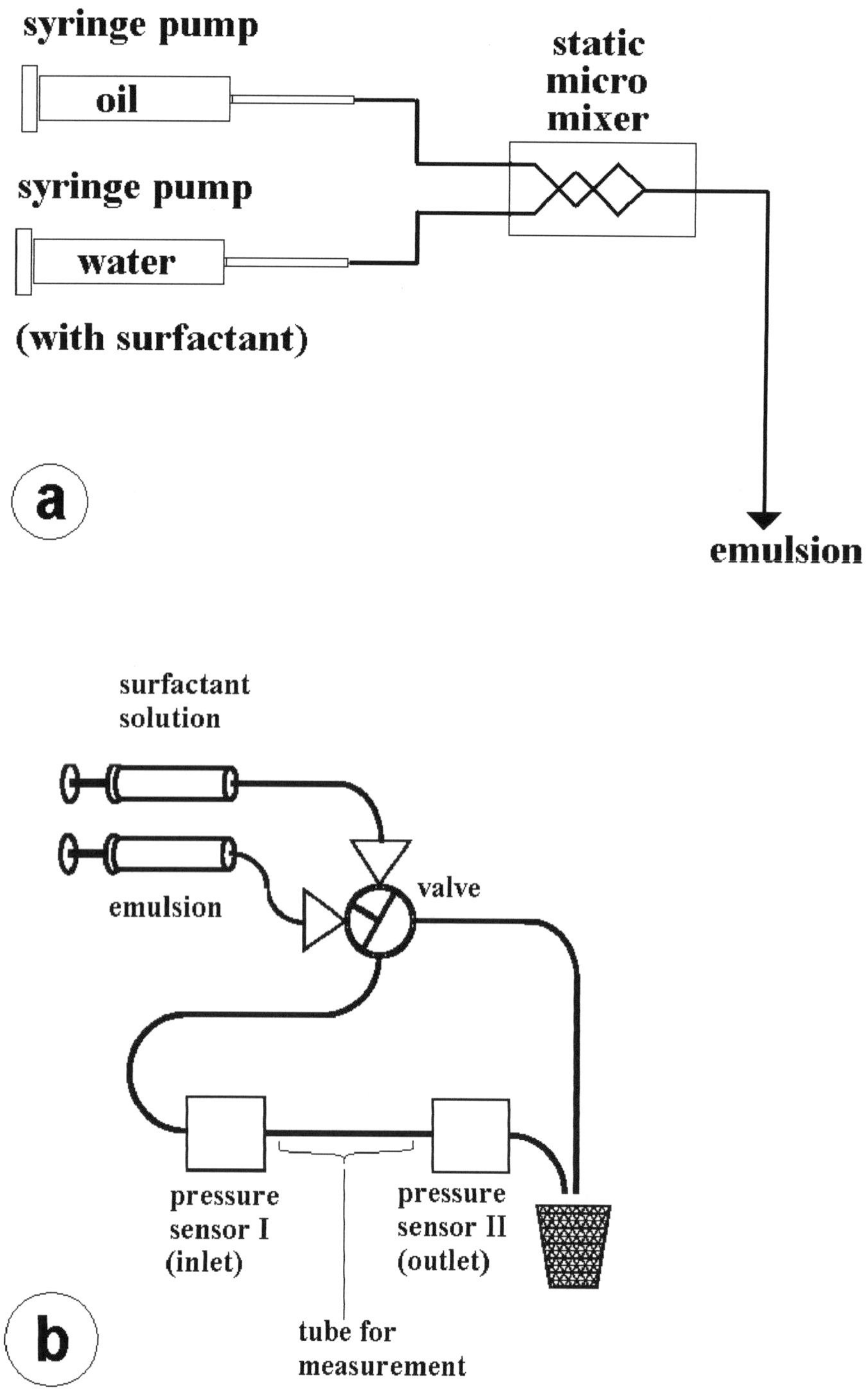

Fig. 1
Experimental arrangements: a) micro fluid arrangement for generation of emulsions by use of a static micro mixer, b) arrangement for pressure drop measurements on streaming emulsions

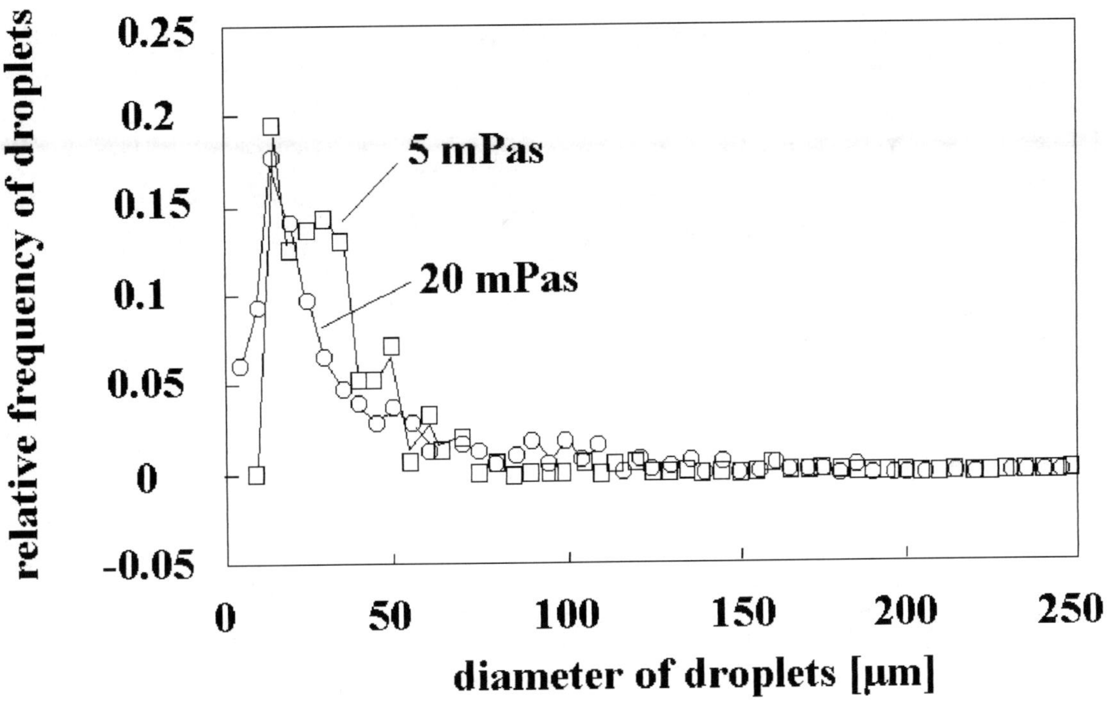

Fig. 2
Size distribution of oil droplets in oil/water emulsions generated by flow-through processes with static micro mixer under presence of SDS for two different silicone oils (high viscosity oil: 20 mPas, low viscosity oil: 5 mPas)

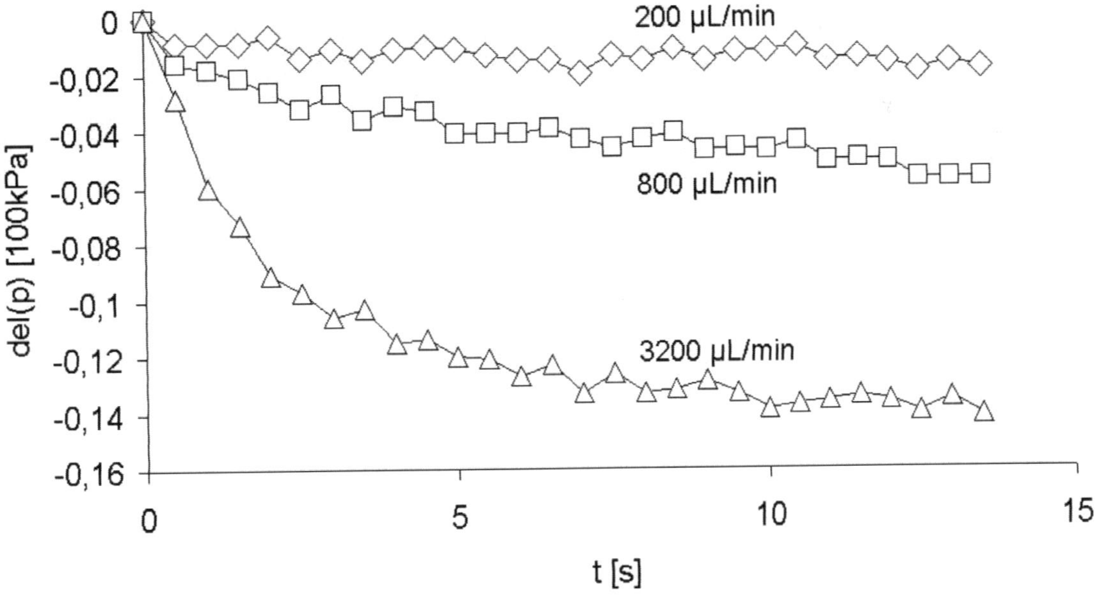

Fig. 3
Evolution of pressure drop after on-switching of emulsion flow through a micro tube for three different flow rates

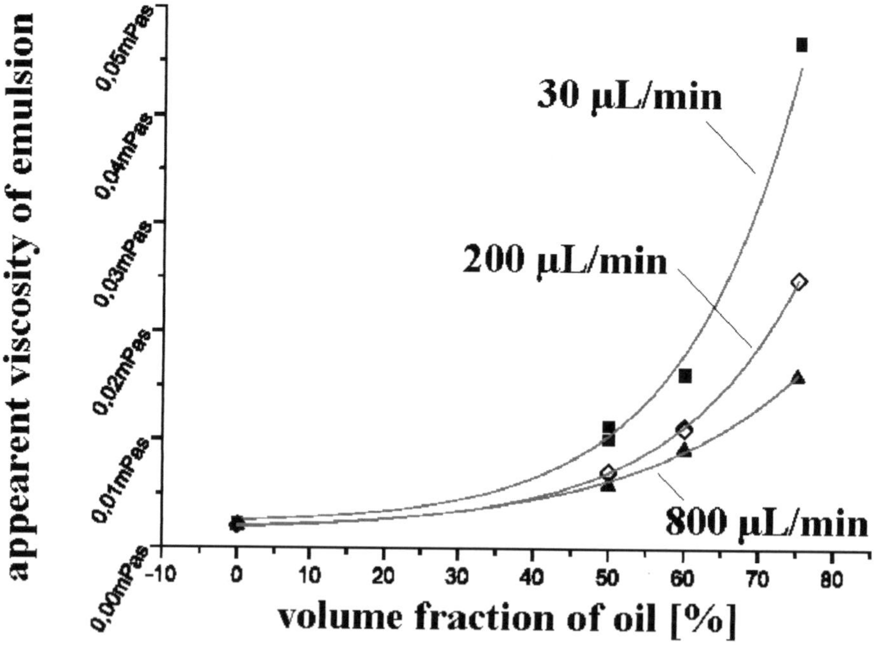

Fig. 4
Dependence of apparent viscosities of oil-in-water emulsions on volume fraction of oil for different flow rates (transport through a micro tube, 0.5 mm internal diameter)

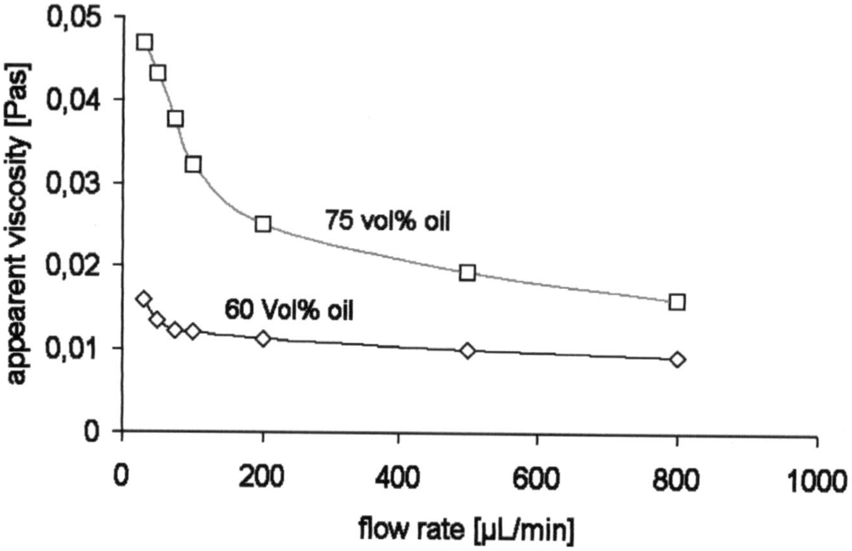

Fig. 5
Decrease of apparent viscosity of oil-in-water emulsions with increasing flow rate for two emulsions with high volume fraction of oil

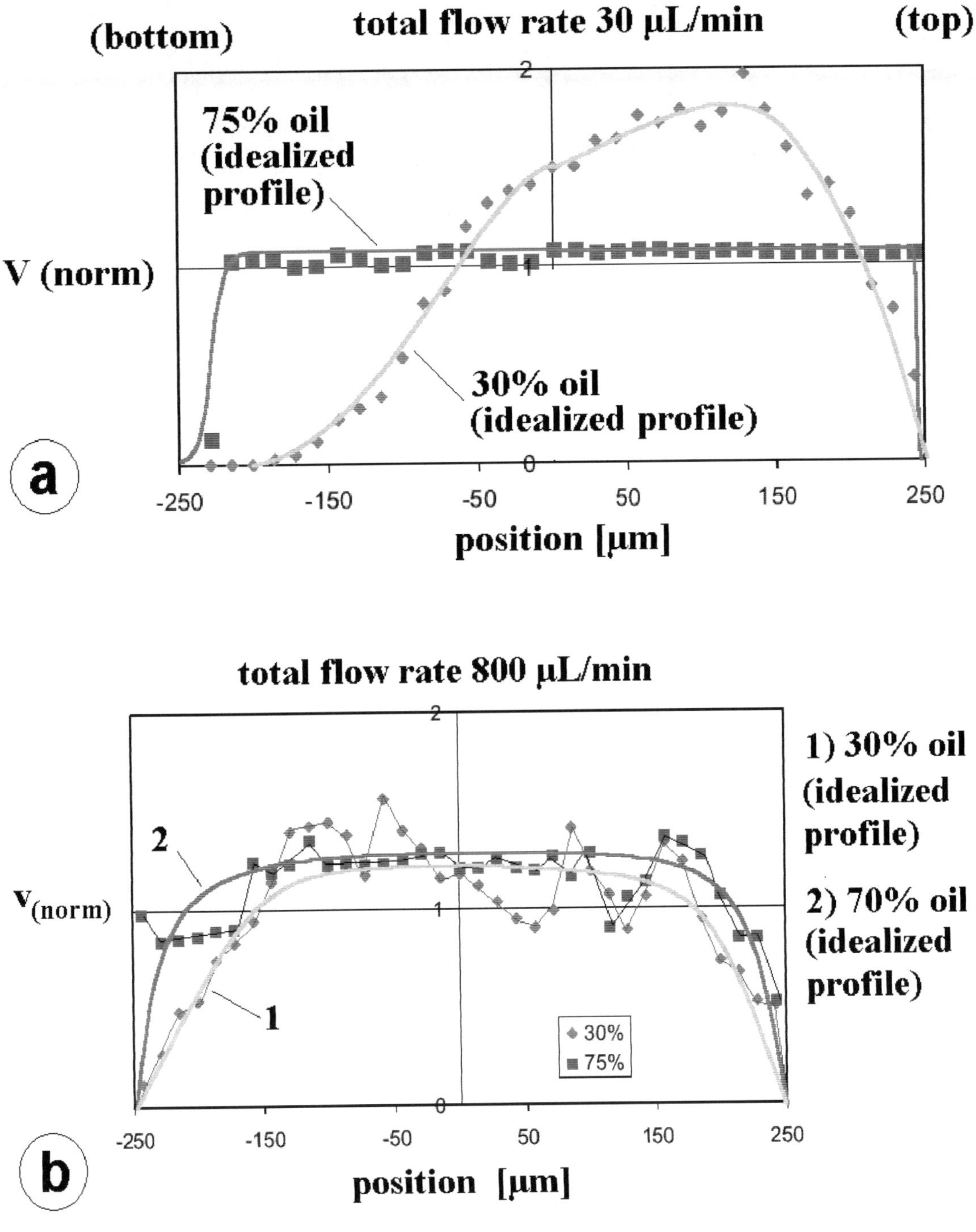

Fig. 6
Velocity distribution of oil droplets inside a lithographically prepared microchannel (500 µm diameter), determined by micro particle velocimetry (asymmetry caused by gravity and horizontal experimental arrangement): a) velocity profiles of emulsions with 30 Vol% and 75 Vol% of oil at a flow rate of 30 µL/min, b) velocity profiles of emulsions with 30 Vol% oil and 70 Vol% oil at a flow rate of 0.8 mL/min

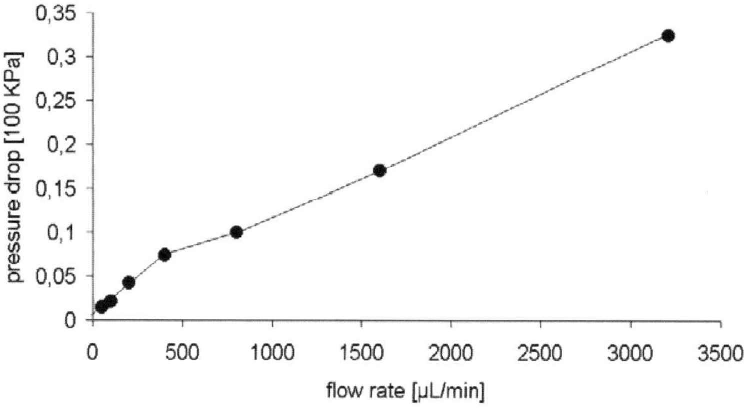

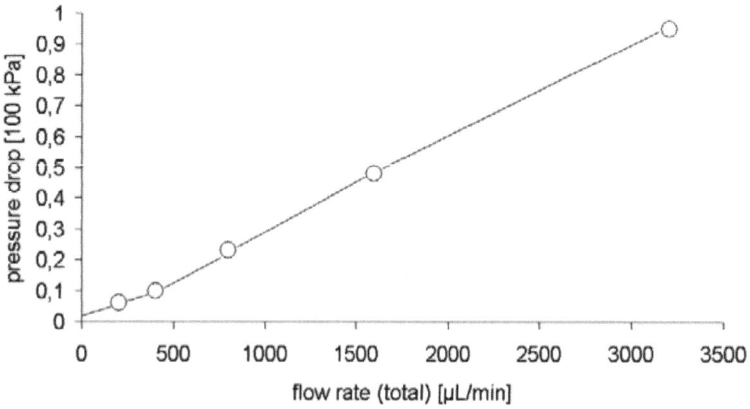

Fig. 7
Dependence of pressure drop on flow rate for streaming emulsions during in-situ formation experiments: a) flow rate ratio: aqueous solution (1% SDS): oil = 1:1, b) flow rate ratio: aqueous solution (1% SDS): oil = 3 : 7

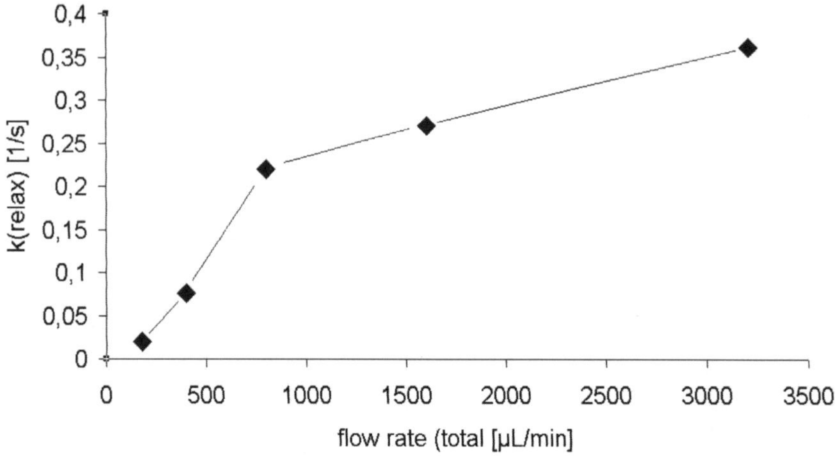

Fig. 8
Dependence of rate of pressure drop relaxation (relaxation of apparent viscosity) after on-switching of flow in the in-situ formation of oil-in-water emulsions (flow rate ratio: aqueous solution (1% SDS): oil = 1:1)

Enhanced Mass Transfer by Liquid–Liquid Slug Flow in Microchannels for Efficient Extraction

Nobuaki Aoki, Teng Hong Khoo, Yoshihito Okubo, and Kazuhiro Mae

Department of Chemical Engineering, Graduate School of Engineering, Kyoto University
Kyoto-daigaku Katsura, Nishikyo-ku, Kyoto 615-8510, Japan
aoki@cheme.kyoto-u.ac.jp

Abstract

The internal circulation liquid–liquid slug flow in microchannels intensifies mass transfer between the two phases. In this work, effectiveness of the slug flow in a narrow channel for the extraction of phenol from dodecane to water was first investigated. Extraction based on slug flow and that based on emulsion generated by a micromixer was compared from the viewpoint of the total operation time for extraction and separation of the two phases. The results showed that the enhanced circulation flow in slugs by increasing flow rate enables rapid extraction. Moreover, large slug length compared with emulsion diameter allows instant separation of the two phases after the extraction. Required time for extraction based on emulsion is shorter than that based on slug flow. However, it takes more than a few hours to separate the two phases due to the small emulsion size (~50 μm). Thus, the extraction based on slug flow is preferable to that based on emulsion from the viewpoint of the total operation time for the entire extraction process. Then, the effects of operating parameters such as ratio of oil phase volume flow rate to total flow rate and outlet channel i.d. were examined. The result showed that appropriate operating condition improves extraction efficiency in slug flow.

Keywords: liquid–liquid slug flow, microchnnel, extraction, mass transfer, internal circulation flow

1. Introduction

Laminar flow in microchannels enables the formation of liquid–liquid slug flow as shown in Fig. 1. Since slugs in the channel move like a periodic plug flow, narrow residence time distribution can be achieved precisely [1,2]. Internal circulation flow in slug derived from the friction between the channel wall and fluid in the slug due to high surface area-to-volume ratio is expected to enhance the performance of mixing, which is mainly driven by molecular diffusion in laminar flow [3,4]. The circulation flow also enhances mass transfer between the slugs of the two phases, since the concentration near the interface is renewed by the flow [5]. The liquid film formed between the reactor wall and droplet phase increase the area of active interface between the two phases, leading to improved mass transfer.

Thus, liquid–liquid slug flow can be used for rapid extraction. In addition, the two phases can be easily separated after the extraction using liquid–liquid slug flow, since the slug size is over several ten times larger than the emulsion size required to a fast extraction. Therefore, extraction in liquid–liquid slug flow is expected to be effective for the entire extraction process.

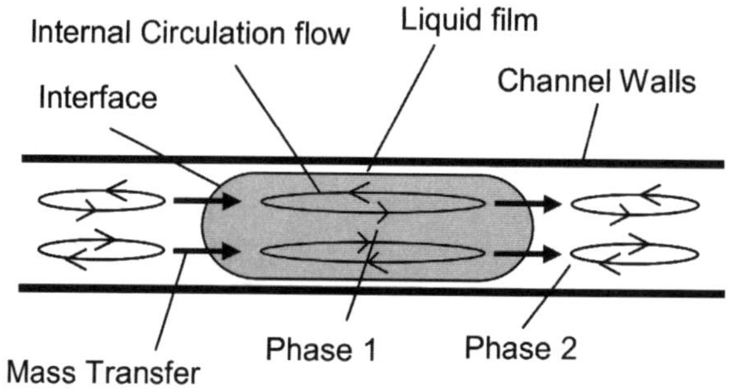

Fig. 1. Internal Circulation flow and mass transfer in slug flow.

In this work, first, effectiveness of the slug flow in a narrow channel for the extraction of phenol from dodecane to water was investigated. For this purpose, we discussed the mass transfer enhancement by the liquid-liquid slug flow by comparing with the stationary contact of the two phases. Extraction based on slug flow and that based on emulsion generated by a micromixer was also compared from the viewpoint of the total operation time for extraction and separation of the two phases. To improve the efficiency of extraction by slug flow and to establish a design method of liquid–liquid slug flow for rapid extraction, we then investigated the relation between operation parameters such as flow rate, ratio of oil phase volume flow rate to total volume flow rate, and outlet channel i.d.

2. Experimental

2.1 Comparison of extraction based on slug flow and that based on emulsion

We first investigated the effectiveness of the slug flow in a microchannel for efficient extraction. To prove the effectiveness of the slug flow, we compared the extraction based on slug and that based on emulsion from the view point of the total operation time for extraction and separation of the two phases.

2.1.1 Slug flow

The experimental setup for extraction based on slug flow is depicted in Fig. 2. Distilled water and 1000 ppm phenol/dodecane were fed to i.d. 1.3 mm, 1/16 inch union tee by using syringe pumps (IC3100, Kd Scientific) through silicon tubes with i.d. 1.0 mm and length 1.0 m. PTFE tubes with i.d. 0.8 mm with lengths calculated based on the residence time and total flow rates were attached to the outlet of the tees. Both the Union Tee and the PTFE tube were immerged in a water bath with T = 25 °C to keep the temperatures constant. The residence time in the outlet tube was fixed at 10 s. The distilled water and phenol/dodecane (Both were purchased from Wako Pure Chemical Industries) were fed with total flow rates of the two phases, V_t = 120, 150, 180, 210, or 240 mL/h, and the ratio of phenol/dodecane flow rate to total flow rate, w_o, was 0.5. The slug lengths of the two phases were measured with a ruler. Both phases were collected in a test tube for 45 s. The concentration of phenol in water phase in the test tube, C_w, was measured with UV–Vis spectrometer (Multispec 1500, Shimadzu). Phenol has its typical peek at 270 nm in distilled water. For comparison, C_w by only contacting dodecane and water phase for 45 s in a test tube was also measured.

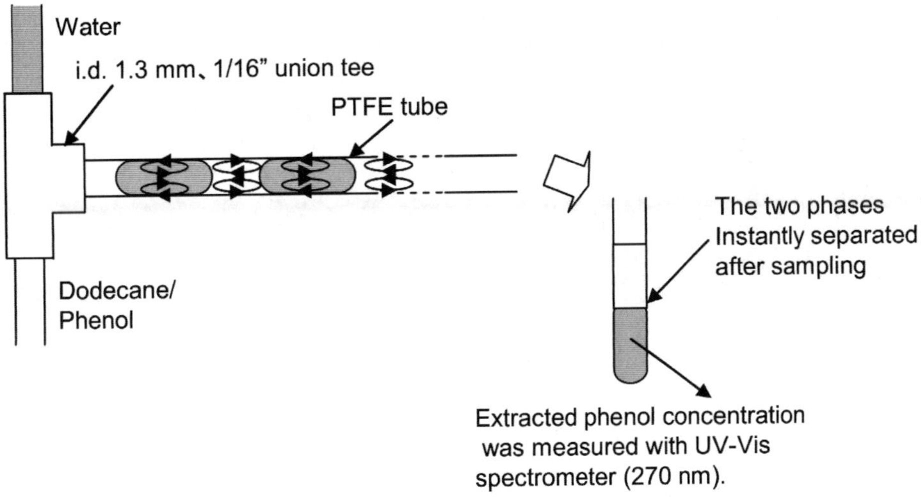

Fig. 2. Experimental set–up for slug flow.

2.1.2 Emulsion

Extraction based on emulsion generated by a micromixer made of aluminum (YM-1, Yamatake) was then carried out. This method also leverages the feature of microchannels, namely, intense interaction between channel wall and fluid due to high surface-to-volume ratio. In YM-1, emulsion is generated through the repeated splitting and recombination of the fluid flow [6]. The internal channel diameter is 400 μm. Distilled water and 1000 ppm phenol–dodecane were fed to the micromixer by using syringe pumps through silicon tubes with i.d. 1.0 mm and length 0.5 m. Instead of PTFE tube, silicon tube with i.d. 1.0 mm and length 0.3 m was attached to the exit of the micromixer. The total flow rate was changed into 40, 60, and 80 mL/min, and w_o was fixed at 0.5. Both phases with emulsion formed were collected in a test tube and centrifugation with spinning speed 500 rpm was carried out for 10 minutes to separate the two phases completely. The concentration of phenol in water phase in the test tube, C_w, was measured with UV–Vis. Fig. 3 shows the micromixer used for the experiment and shows the samples of emulsion. The diameter of emulsion was also measured using a Particle Size Distribution Analyser (LA-920, Horiba).

(a) YM-1 (aluminum) (b) Emulsion

Fig. 3. Micromixer and samples of emulsion.

2.2 Effects of operating conditions

2.2.1 Flow rate ratio of the two phases

In order to investigate the effect of flow rate ratio of the two phases on the extraction efficiency of liquid–liquid slug flow, experiments were carried out using the same method as described in the previous section with various value of w_o = 0.3, 0.5, 0.7, or 0.9. In these experiments, temperature and V_t were fixed at 25 °C and 180 mL/h, respectively. The same union tee and PTFE tube was used.

2.2.2 Outlet channel i.d.

We also examined the relation between extraction efficiency and PTFE tube i.d. The PTFE tube i.d., *D*, was changed into 0.5, 0.8, or 1.0 mm. Fixed conditions were as follows: V_t = 180 mL/h; w_o = 0.5; and T = 25°C.

3. Results and Discussion

3.1 Comparison of extraction based on slug flow and that based on emulsion

Fig. 4 shows the effect of V_t on C_w and average slug length of the two phases. The results indicate that the increase in total flow rate improved the concentration of extracted phenol. This is due to the circulation rate in slugs is enhance with increasing V_t. No extraction occurred when the phases contacted only in the test tube (V_t = 0 mL/h and without circulation flow). This also indicates that mass transfer occurs only in the outlet tube when slug flow is formed. Thus, rapid internal circulation flow enhances extraction efficiency. With regard to required time for two phase separation, the water and oil phase were separated right after sampling in the test tube. For the extraction based on slug flow, the total operation time for extraction and two phase separation is thus approximately equal to the time for extraction.

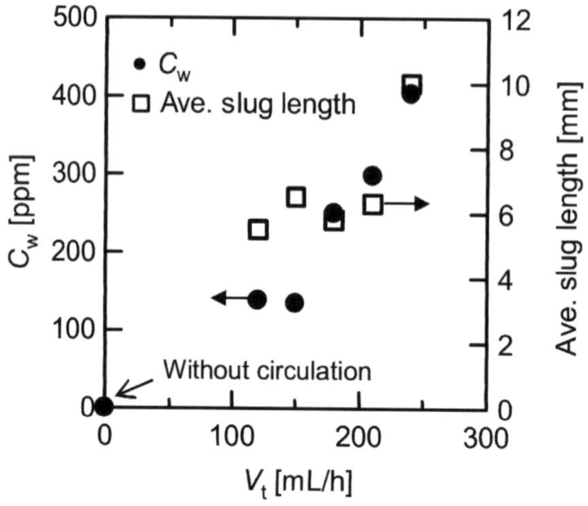

Fig. 4. C_w and slug length by the slug flow.

Fig. 5 shows the extracted concentration of phenol and the emulsion diameter. The extracted concentration of phenol was much higher than that obtained by slug flow. At high V_t, the value of C_w is close to an equilibrium concentration (C_w^* = 9.8 × 10^2 ppm). Meanwhile the emulsion diameters were about 50 μm. The surface area was extremely large and rapid extraction was enabled. Therefore, the rapid extraction occurred in a very short time. However, before the measurement of extraction concentration of phenol, centrifugation was necessary to separate the two phases completely because of the small size of emulsion. This extraction method enables more rapid extraction than that based on slug flow. However, without any operation, this method requires more than a few hours to separate the two phases after sampling because of the much smaller emulsion sizes than slug sizes shown in Fig. 4. Thus, the total operation time for the extraction and separation of the two phases based on emulsion is longer than that based on slug flow.

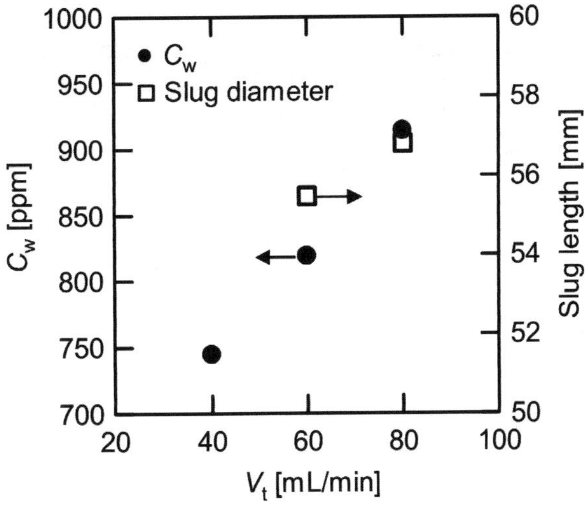

Fig. 5. C_w and emulsion diameter by the micromixer.

Table 1 summarizes the scale of operation time of extraction and the two phase separation for the two extraction methods. Though the extraction time for slug flow is longer than that for emulsion, the two phase separation time for slug flow is nearly negligible. Thus, extraction and separation of the two phases can be performed in a short time in one continuous channel with branched hydrophobic and hydrophilic outlets. A continuous system that a phase separator is connected right after an extractor has been developed [7].

Table 1. Comparison of operation time for extraction and two phase separation

Extraction base on	Extractoin time	Two phase separation time
slug flow	~10 s	instantaneous
emulsion	~1 s	~hour

3.2 Effects of operating conditions

3.2.1 Flow rate ratio of the two phases

Fig. 6 shows the extraction ratio and the slug length as a function of w_o. Extraction ratio is obtained by

$$\text{Extraction ratio [-]} = \frac{\text{Phenol transferred to water [mol]}}{\text{Phenol provided from dodecane [mol]}} = \frac{C_w \text{ [ppm]}}{\text{initial concentration of phenol in dodecane [ppm]}} \times \frac{1-w_o}{w_o} \quad (1)$$

As a result, the maximum extraction ratio was obtained at w_o = 0.7. This is partly due to the small average slug length that improves the circulation rate. However, this value is larger than that for taking the shortest average slug length, that is, w_o = 0.5. The length of slug of distilled water decreases with increasing w_o. At high w_o, the circulation time in water phase is short and the mass transfer in the phase is enhanced. This result indicates that the internal circulation in the water phase dominates the extraction rate.

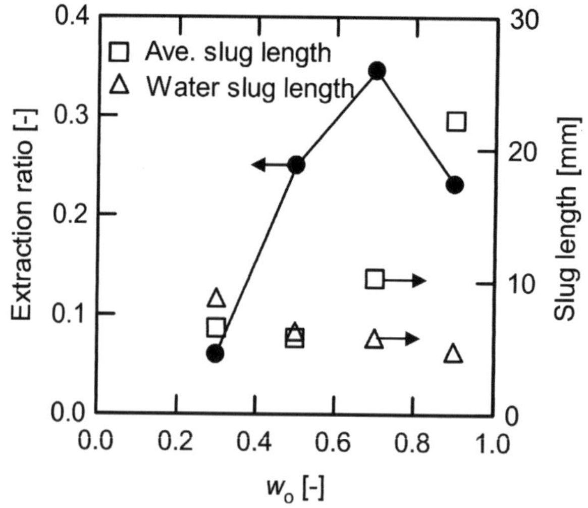

Fig. 6. Extraction ratio and slug length with various w_o at T = 25 °C, V_t = 180 mL/h.

3.2.2 Outlet channel i.d.

The effect of PTFE tube i.d. on extraction ratio of phenol and slug length is plotted in Fig. 7. When D = 0.8 mm, the extraction ratio took the highest value. The increase in PTFE tube i.d. decreased the slug lengths of both phases. The decrease in slug length enhances the circulation in slugs. However, the traveling velocity of the slugs in the PTFE tube decreases with increasing tube i.d. The internal circulation rate decreased with reducing traveling velocity [3]. As a result, enlarging the PTFE tube i.d. also has an effect of reducing the internal circulation flow. Thus, there is a trade-off between the effects of channel i.d. on extraction efficiency.

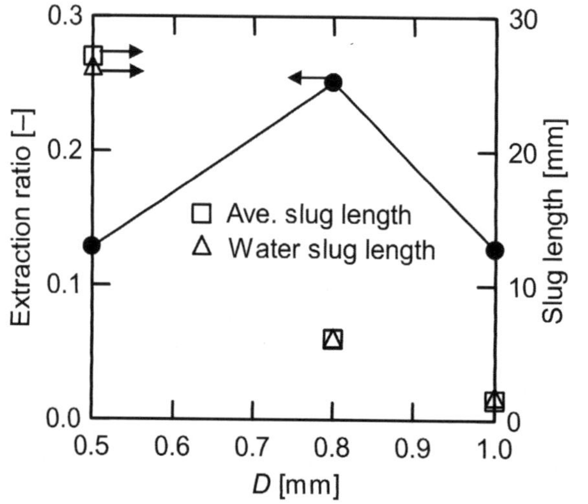

Fig. 7. Extraction ratio of various i.d. PTFE tubes.

4. Conclusions

Effectiveness of liquid–liquid slug flow for rapid extraction has been investigated. The results showed that the circulation flow in slugs enables rapid extraction. The study with various operating conditions indicates that appropriate conditions further improve the efficiency of extraction by slug flow. Moreover, large slug size compared with emulsion size allows the two phases to be separated instantly after the extraction. These advantages suggest that the slug flow has a potential for high throughput extraction by increasing flow rate and parallelizing channels and that the extraction and separation of the two phases can be performed in a short time in one continuous channel with branched hydrophobic and hydrophilic outlets.

Acknowledgment

This research was partially supported by the New Energy and Industrial Technology Development Organization (NEDO), Project of Development of Microspace and Nanospace Reaction Environment Technology for Functional Materials.

References

[1] Song, H., J. D. Tice and R. F. Ismagilov; "A Microfluidic System for Controlling Reaction Networks in Time," *Angew. Chem. Int. Ed.*, **42**, 768–772 (2003).

[2] Trachsel, F., A. Günther, S. Khan and K. F. Jensen; "Measurement of Residence Time Ddistribution in Microfluidic Systems," *Chem. Eng. Sci.*, **60**, 5729–5737 (2005).

[3] Tanthapanichakoon, W. , N. Aoki, K. Matsuyama and K. Mae; "Design of Mixing in Microfluidic Liquid Slugs Based on a New Dimensionless Number for Precise Reaction and Mixing Operations," *Chem. Eng. Sci.*, **61**, 4220–4232 (2006).

[4] Tanthapanichakoon, W. K. Matsuyama, N. Aoki and K. Mae; "Design of Microfluidic Slug Mixing Based on the Correlation between a Dimensionless Mixing Rate and a Modified Peclet Number," *Chem. Eng. Sci.*, **61**, 7386–7392 (2006).

[5] Harries, N., J. R. Burns, D. A. Barrow and C. Ramshaw; "A Numerical Model for Segmented Flow in Micro-reactor," *Int. J. Heat Mass Transfer*, **46**, 3313–3322 (2003).

[6] Mae, K., T. Maki, I. Hasegawa, U. Eto, Y. Mizutani and N. Honda; "Development of a New Micromixer Based on Split/Recombination for Mass Production and its Application to Soap Free Emulsifier," *Chem. Eng. J.*, **101**, 31–38 (2004).

[7] Kralj, J. G., H. R. Sahoo and K. F. Jensen; "Integrated Continuous Microfluidic Liquid–Liquid Extraction," *Lab Chip*, **7**, 256–263 (2007).

Flow rate variation in two step multi-phase microsynthesis of nanomaterials affecting their catalytic activity of oxidation

J.Michael Köhler, Lahbib Abhamane, Katrin Ritter and P. Alexander Groß
Technical University of Ilmenau, Institute for Micro and Nanotechnologies/Institute for Physics

correspondence:
Prof. Dr. M.Köhler, Techn. Univ., D-98693- Ilmenau, Inst. f. Micro and Nanotechnologie, Inst f. Physics, Germany, phone 03677-693700, michael.koehler@tu-ilmenau.de,

Abstract

The effect of flow rates during the micro flow-through synthesis of gold-containing binary nanoparticles and their catalytic activity were studied. Therefore, a micro fluid system with 3 injectors and mixing sections was applied. A multiphase flow system was applied in order to realize a plug-like liquid transport, and therefore a narrow residence time distribution between the used second and third injector positions. The catalytic activity of the obtained nanomaterials was tested by the oxidative degeneration of dyes (rhodamine B, bromphenolblue) in aqueous solution in the presence of potassium peroxodisulphate. The bleaching rates of reaction suspensions containing about 0.1 weight percent of catalyst increased up to factors of about 15 in comparison with solutions without catalyst. The factor of reaction activity enhancement was dependent on the composition of educt solutions as well as on the flow rate during the particle preparation.

Keywords

nanoparticles, gold, silver, iron, micro flow-through synthesis, catalytic oxidation, catalytic activity

1. Introduction

Metal nanoparticles of different composition are of interest for a wide spectrum of applications and particularly for catalytic purposes [1-3]. Their high effective surface make them suited for catalytic flow-through processes in micro reactors with short residence times. Gold and silver nanoparticles are particularly important for selective partial oxidation of organic compounds. The catalytic activity and specificity are influenced by the distribution of sizes and compositions within a population of nanoparticles. The quality, and in particular the uniformity of catalytic nanomaterial can be improved if the nanoparticles are prepared by using a micro flow-through precipitation process. Previous investigations showed that the properties of colloidal solutions of Au/Ag particles obtained by a miniaturized flow-through process are strongly dependent on flow rates. The different qualities are reflected by strong variations in optical spectra.

Small particles are of particular interest for catalytic applications due to their high specific surface. There are nanoparticles showing the same catalytic quality like the corresponding bulk materials. For example, oxid particles for proton transfer processes, palladium and platinum particles for hydrogen transfer or silver particles for oxidation reactions. Nanomaterials are more active in these reactions in cause of the high number of surface atoms. Beside this quantitative effect, nanoparticles are carrying frequently surface atoms in a special local environment and a special chemical activity. In contrast, there are nanoparticles with catalytic activity different from the chemical properties of the bulk materials. Gold nanoparticles with high selectivity for partial oxidation of olefines are an example for an unique catalytic effect of nano-sized metallic particles [4, 5].

Nanoparticles for catalytic applications can be produced by gas phase or by liquid phase processes. The quality of particles is strongly determined from the reaction conditions. In particular, the local distribution of reactants is important for the rate of nucleation and the growth rate of nanoparticles. Both processes must be controlled strictly, if a homogenous population of nanoparticles should be formed. A narrow distribution in shape and size of nanoparticles as well as a homogenous chemical composition are preconditions for equal catalytic activity of the particles.

Micro fluidic arrangements and micro reactors are particularly suited for the mixing of reactants in flow-through processes under well controlled conditions. The prevalence of laminar flow causes a well controllable convection inside microchannels and can support constant streaming conditions and steep and stationary concentration gradients inside the reaction system. So, static micro mixers can be very advantageously used for fast and efficient mixing in micro flow-through processes. Micro flow-through systems can be applied for the preparation of inorganic dielectric particles [5-7], for polymer nanoparticles [8] as well as for the preparation of semiconductor [9-11] and metal [12] nanoparticles. The generation of nanoparticles in micro flow-through processes can occur in homogenous phase of reactant solutions or by application of multiphase flow and nanoparticle formation inside fluid segments [13]. It was shown, that the quality of nanoparticles depends not only on the type, concentration and ratio of reactants, pH and other chemical parameters, but can also be influenced by the flow conditions. So, it was found that the mean size of Au-nanoparticles obtained by reduction of tetrachloroaurate by ascorbic acid at room temperature was decreased with increasing flow rate. In addition, the width of size distribution could be reduced with increasing flow under constant reactant ratio [14-16]. The generation of gold nanorods succeeded under the particular conditions of a micro flow and high surfactant concentrations[17,18].

The convection inside the microfluidic system is of particular importance in two and multistep mixing processes. In this case, not only the single mixing processes but also the residence time between the mixing steps depends on the flow rate. So, the optical properties of colloidal product solutions obtained in a two step flow-through mixing process by reduction of gold and silver compounds was very strongly dependent on the flow rate. This effect was discussed in relation to the distribution of Au and Ag between the core and the shell of nanoparticles [19]. This distribution should not only be important for the optical but also for the catalytic properties of nanoparticles. Therefore, here the preparation of catalytic active nanoparticles was studied in dependence on flow conditions and the catalytic activity of the obtained material was tested using a simple chemical bleaching reaction.

2. Experimental

A microfluidic system completely based on PTFE ware was used for the generation of nanoprecipitates and colloidal solutions. The system consisted on 4 fluid actuators (syringe pumps), three T-type injectors, two tube coils for mixing and an outlet for colleting the product (fig. 1).

At the first injector, fluid segments are formed from aqueous solution of reducing agents (sodium borohydride/ascorbic acid) injected into inert carrier liquid (PP9, perfluorinated alkane). We used asorbic acid, sodium borohydride as reducing agents. Some experiments were carried out by mixtures of both chemicals. In some cases, sodium dodecylsulphate (SDS, surfactant) was applied in addition. At the second injector, a tetrachloroaurate solution (5mM) in water was injected into the aqueous liquid segments of reducing agent in order to start reductive nucleation. The length of the following residence multi loop (20 cm) determined the time between starting of metal nucleation and addition of the second metallic component. At the third injector, salt solutions of the second metal component (silver, silver/gold mixture, palladium, iron) were added to the aqueous reaction mixtures in the fluid segments. Silver ions were applied in form of a 50 mM solution of silver nitrate or in mixtures of silver nitrate and tetrachloroauric acid, in general.

PTFE ware was used for tubes (0.5 mm diameter) and for injectors. The flow rates were varied between 50 and 5000 µL/min resulting in delay times between the second and the third injector between approximately 50 ms and five seconds.

The obtained nanoparticles were precipitated and cleaned by centrifugation and washing.

The oxidation of a dye (rhodamin G) by potassium peroxodisulphate in aqueous solution was chosen as an example for the catalytic activity of nanomaterials. Therefore the nano precipitates were suspended in a freshly prepared aqueous solution of the dye and the oxidizing agent. The reaction was monitored by taking absorption spectra of reaction mixture over time. Between the optical measurements, the solutions were mechanically actuated in order to enhance the interaction between reaction mixture and catalyst by convection. The enhancement of bleaching rate of dye in presence of nanoparticles in comparison with the nanoparticle-free solution was used for the estimation of the catalytic activity.

3. Results and discussion

The nanoparticles were obtained either in form of coloured colloidal solutions or as grey precipitates. For catalytic tests, all product solutions were transferred into precipitates by centrifugation, in order to wash and dry them. Fig. 2 gives some examples of products. The precipitates shows a very high internal roughness and high internal surface, in general. All catalyst materials possess a very high porosity with pore sizes to far below 50 nm (fig. 2, a,b). In case of colloidal solutions of binary nanomaterials, core shell particles (fig. 2c) could be observed in some cases (fig. 2c). Small particles with high homogeneity of size were found in reduction experiments with tetrachloroauric acid using sodium borohydride for reduction (Fig. 2d). Reduction of gold nanoparticles by ascorbic acid (0.1 M) at room temperature supply nanoparticles with diameters in the range of about 12 .. 40 nm. Much smaller nanoparticles were obtained at same conditions if a borohydride solution is used instead of the ascorbic acid (fig. 3a). The mean size of nanoparticles can be shifted to some extend by changing the concentration or the flow rate ratio of educts and water, respectively (fig. 3b). The size of Au nanoparticles formed by reduction with ascorbic acid can be shifted by pH, solution composition and flow rate [Wagner]. Fig. 4 show the influence of the flow rate of a sodiumhydrogenesulphate solution at constant flow rates of tetrachloroaurate, reducing agent and further additives. The size can be changed between about 12 and 42 nm (fig. 4a) The shape of the function of nanoparticle size on pH is only moderately influenced by the flow rate of ascorbic acid (fig. 4b) and polyvinylpyrrolidone solution (fig. 4c).

The nanoparticle preparation was always started by the generation of gold cores (first mixing step). After the first residence loop, a solution of an other metal salt or salt mixture was added.

Organic dyes like rhodamine G were bleached by strong oxidizing agents. Potassium peroxodisulphate possess a very high redox potential and is able to destroy the organic molecule by oxidative degradation. The observed decrease in light absorption of aqueous solution of dye after addition of the oxidizing agent is due to this redox reaction. The process is temperature-dependent and can be enhanced by the addition of catalysts. A typical example for the change in optical spectra after mixing of an Au/Ag-catalyst-containing tube with a mixture of dye solution and oxidizing agent is shown in fig. 5. Three mg of catalyst in a 1 ml-tube are sufficient for getting a complete bleaching of solution in less than 40 minutes at room temperature.

The effect of metal particles on the bleaching rate is considerably. In nearly all cases, a increase in bleaching rate of a factor of three or more was observed. Only a moderate effect of the composition of reducing agent on catalytic activity was found in the preparation of Au/Ag nanoparticles with sodiumborohydrate. Similar bleaching rates are obtained for nanoparticles generated by a 50 mM $NaBH_4$ solution with addition of 1 % SDS (fig. 6a) and by a 50 mM $NaBH_4$ solution containing 50 mM ascorbic acid (fig. 6b). The preparation was carried out at three different flow rates but under same flow rate ratios for all reactant solutions and the carrier liquid. The order in differences in bleaching rates of the products (fig. 6a) obtained at different flow rates are small, but are reproduced for the second experiment group with modified composition of the reducing agent (6c). This speaks for a certain effect of flow rate on the catalytic activity of the prepared materials. An other set of experiments hints to the existence of an optimum in the flow rate for getting Au/Ag nanoparticles with highest bleaching activity (fig. 6c).

The catalytic activity of silver on oxidation is well known. In contrast, pure gold particles show nearly know catalytic activity in the test reaction used here. But, the colloidal gold solutions affect the formation of silver particles strongly. The act as catalyst in the growth of silver from silver salt solution. Small silver particles and very fine structured precipitates are obtained, if gold cores are used for metal nucleation. So, the catalytic activity of silver is supported by the effect of gold cores on the formation of metallic silver.

Beside silver nitrate, also iron and palladium salts were tested for their effect on the dye bleaching after reductive preparation of nanostructured precipitates in the presence of gold nanoparticles. A significant increase of bleaching rate of rhodamine G was found in case of palladium as well as in case of iron. The effect on catalytic activity is higher with palladium-containing particles (fig. 7a). In contrast, precipitates obtained by reduction of $FeCl_3$ (50 mM) by sodium borohydride in the presence of Au nanoparticles are marked by a higher effect of flow rate during nanoparticle preparation (fig. 7b).

The distribution of metallic silver and the presence of silver at the surface of the nanoparticles is obviously deciding for the bleaching activity. This fact is supported by the result from experiments with different order in application of gold and silver during the 2-step nanoparticle generation (reduction by borohydride). The lowest enhancement of oxidation rate was found, if a mixture of tetrachloroaurate and silver nitrate is applied in the second reaction step (Fig. 8, graph b). The highest activity was found if the core formation was executed from a mixture of tetrachloroaurate and silver nitrate solution in a first step, followed by the deposition of silver after the passage of the residence tube (Fig. 8, graph d). In this case, the reaction is more than ten times accelerated in comparison with the nanoparticle-free reactant mixture.

4. Conclusions

Micro fluidic arrangements including injectors, mixing and residence loops are suited for two and multistep processes in generation of binary metallic particles for catalytic applications. The formation of silver, iron and palladium-containing nanoparticles and nanostructured precipitates succeeded by formation of gold nanocores at first. The application of sodium borohydride for reduction of gold from tetrachlororate solutions at room temperature leads to cores in the range of about 4..6 nm. The obtained colloidal solutions can be taken for deposition of further metals by metal-catalysed metal reduction. So binary nanoparticles containing silver, palladium or iron can be prepared which possess a significant catalytic activity of oxidation. Here, the enhancement of bleaching rate of an organic dye by potassium peroxodisulphate was chosen as test reaction.

The catalytic activity is not only dependent on the used metals but is also influenced by the reaction procedure. Certain differences in bleaching rates were found in relation to the order of reactant application and on flow rate during the nanoparticle preparation. The investigations demonstrate the applicability of microreaction technology for the preparation of catalysts for heterogeneous phase processes and the possibility of optimize the properties of catalytic nanoparticles by controlling two and multistep processes.

Acknowledgement

We thank F. Möller and S. Schneider for carrying out experiments and for technical support. Financial support by the German Environmental Foundation (DBU) is gratefully acknowledged.

References

[1] W. J. Parak, L. Manna, F. Ch. Simmel, D. Gerion, P. Alivisatos, in: Nanoparticles (ed. by G. Schmid (Weinheim 2004), 4-49

[2] U. Simon, Advanced Materials 1999, 10, 1487

[3] U. Simon, in: Nanoparticles (ed. by G. Schmid (Weinheim 2004), 328-362

[4] M. Haruta, Nature 2005, 437, 1098

[5] M.D. Hughes et al: Nature 2005, 437, 1132

[6] R. Schenk, M. Donnet, V. Hessel, Ch. Hofmann, N. Jongen, H. Löwe: IMRET 5, Strasbourg, France, May 2001, 489

[6] V. Hessel, H. Löwe, Chemie Ingenieur Technik 2002, 74, 17 und 185

[7] R. Schenk, V. Hessel, N. Jongen, V. Buscaglia, S. Guillemet-Fritsch, A.-G. Jones, Nanopowders Produced using microreactors, in: Encyclopedia of Nanoscience and Nanotechnology (ed. by H.S. Nalwa) 2004, Vol 10

[8] P.M. Günther, J. Wagner, G.A. Groß, J.M. Köhler, 9th Internatl. Conf. On Miniaturized Systems :-TAS (Boston, October 9-13) 2005, 918-920

[9] H. Nakamura, Y. Yamaguchi, M. Miyazaki, M. Uehara, H. Maeda. P. Mulvaney (2002) Chemistry Letters: 1072-1073

[10] H. Nakamura, A. Tashiro, Y. Yamaguchi. M. Miyazaki, T. Watari, H. Shimizu, H. Maeda (2004), Lab on a Chip 4: 237-240

[11] Chan E M, Mathies R A, Alivisatos A P (2003) Nano Letters 3: 199-201

[12] J. Wagner, T. Kirner, G. Mayer, J. Albert, J.M. Köhler, Chem. Eng. J. 2004, 101, 251

[13] I. Shestopalov, J. D. Tice, R.F. Ismagilov, Lab Chip 2004, 4, 316-321

[14] J.Wagner, J. M. Köhler, Nano Letters 2005, 5, 685

[15] J. Wagner, T.R. Tshikhudo, J.M. Köhler, 9th Internatl. Conf. on Microreactor Technology IMRET 9 (Potsdam, 6.-8.Sept. 2006), 128

[16] J.M. Köhler, U. Hübner, J. Wagner, M. Held: AICHE Spring National Meeting/5th world congress on particle technology (Orlando, USA, April 2006), paper 98c

[17] C. Sönnichsen, C.; A.P. Alivisatos: Nanoletters 2005, 5, 301-304

[18] C. Sönnichsen, C.;A.P. Alivisatos: Biophys. J. 88 (2005), 364 A - 365 A

[19] M. Moskovitch, I. Srnouvá-Sloufová, B.J. Vlckova, Chem. Phys. 2002, 116, 10435-10446

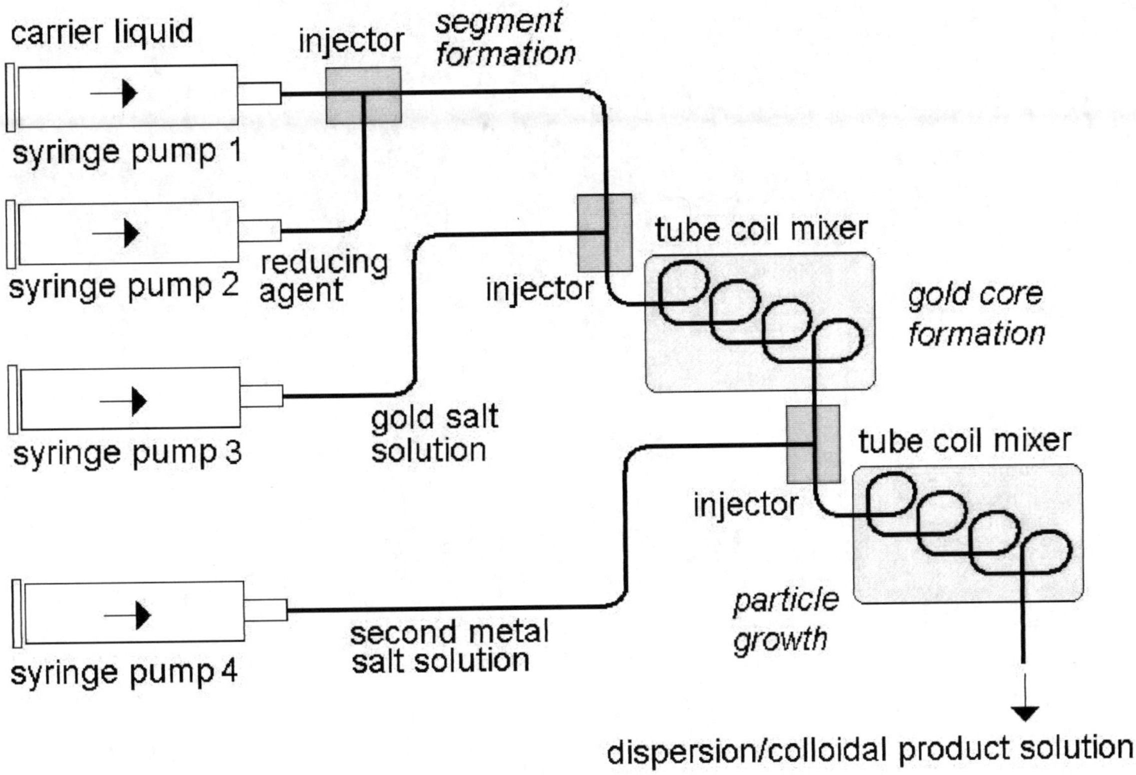

Fig. 1
Micro fluid arrangement for nanoparticle generation in a two step flow-through process

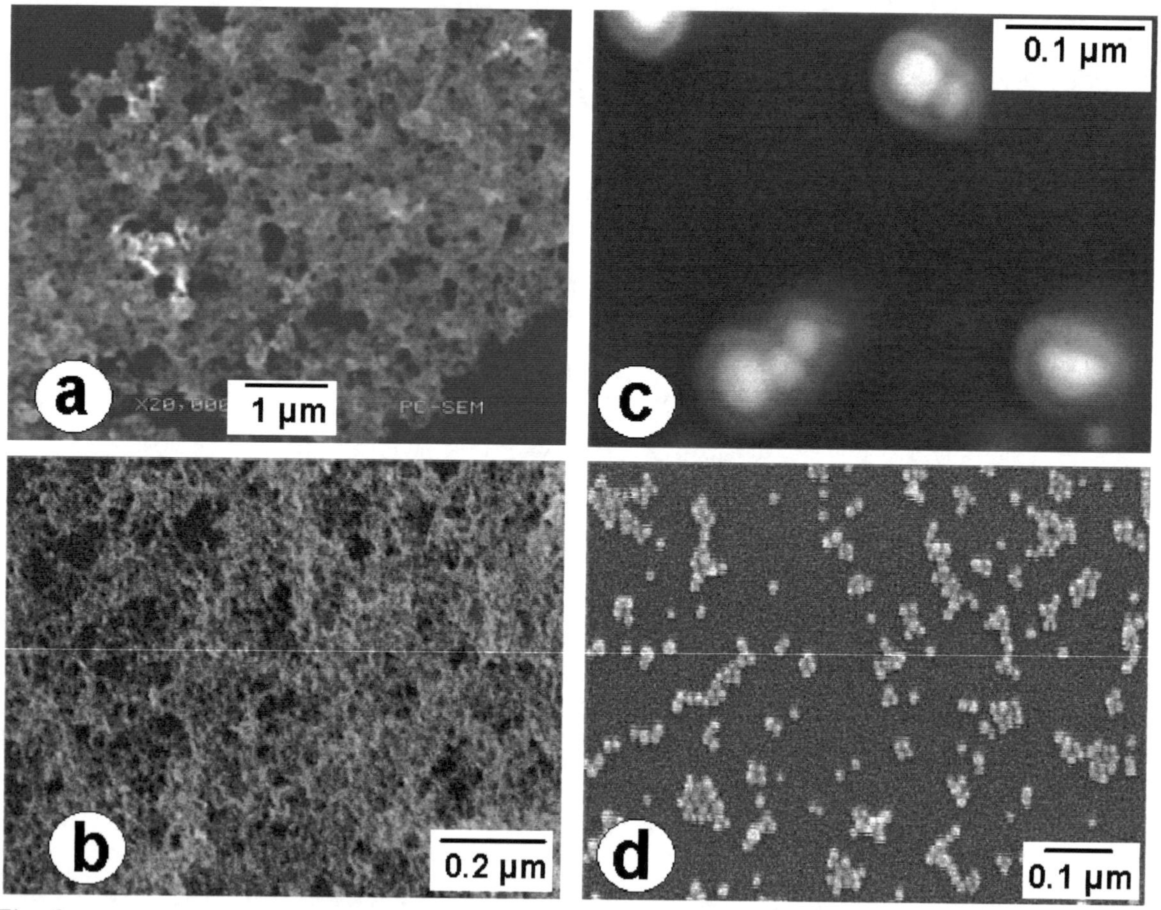

Fig. 2

Examples for Au/Ag nanoprecipitates and nanoparticles for catalytic tests: a,b) precipitates obtained by reduction Au and Ag ions containing solution by sodium borohydride, c) core/shell Au/Ag nanoparticles obtained by a two-step process, d) Au nanoparticles obtained as colloidal solution by borohydride reduction

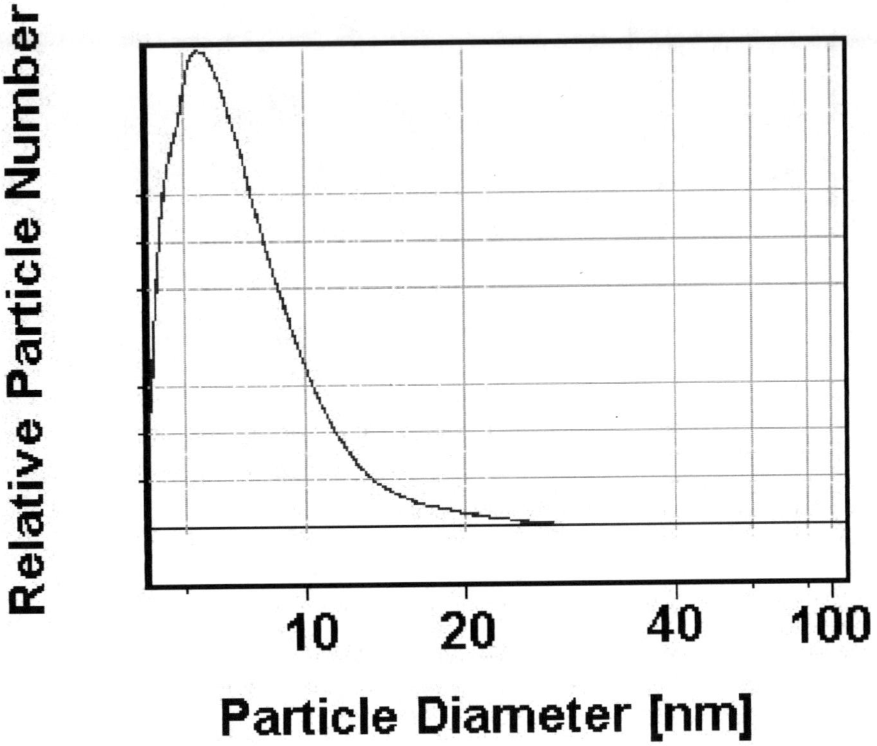
Relative Particle Number
10
20
40
100
Particle Diameter [nm]

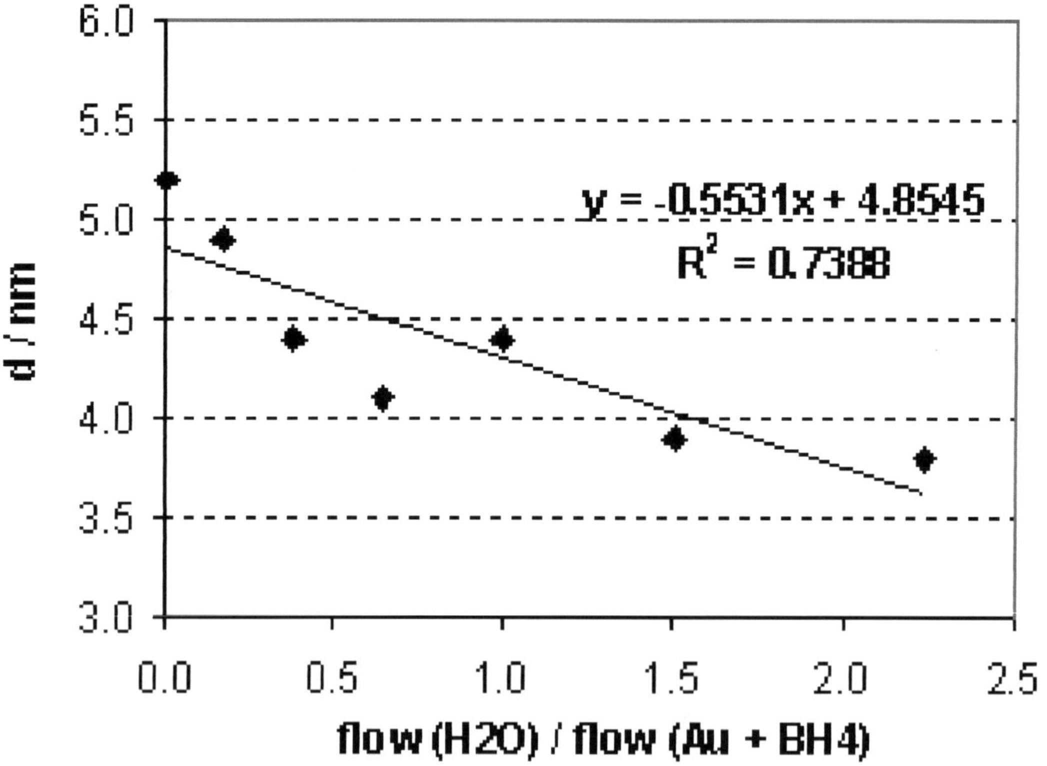

Fig. 3
Preparation of small Au nanoparticles by borohydride reduction with $HAuCl_4$ at room temperature: a) particle size spectrum (obtained by centrifugal sedimentation spectroscopy), b) dependence of mean particle size on concentration of reactants (flow rate ratio between reactants and water in a multi-mixing process)

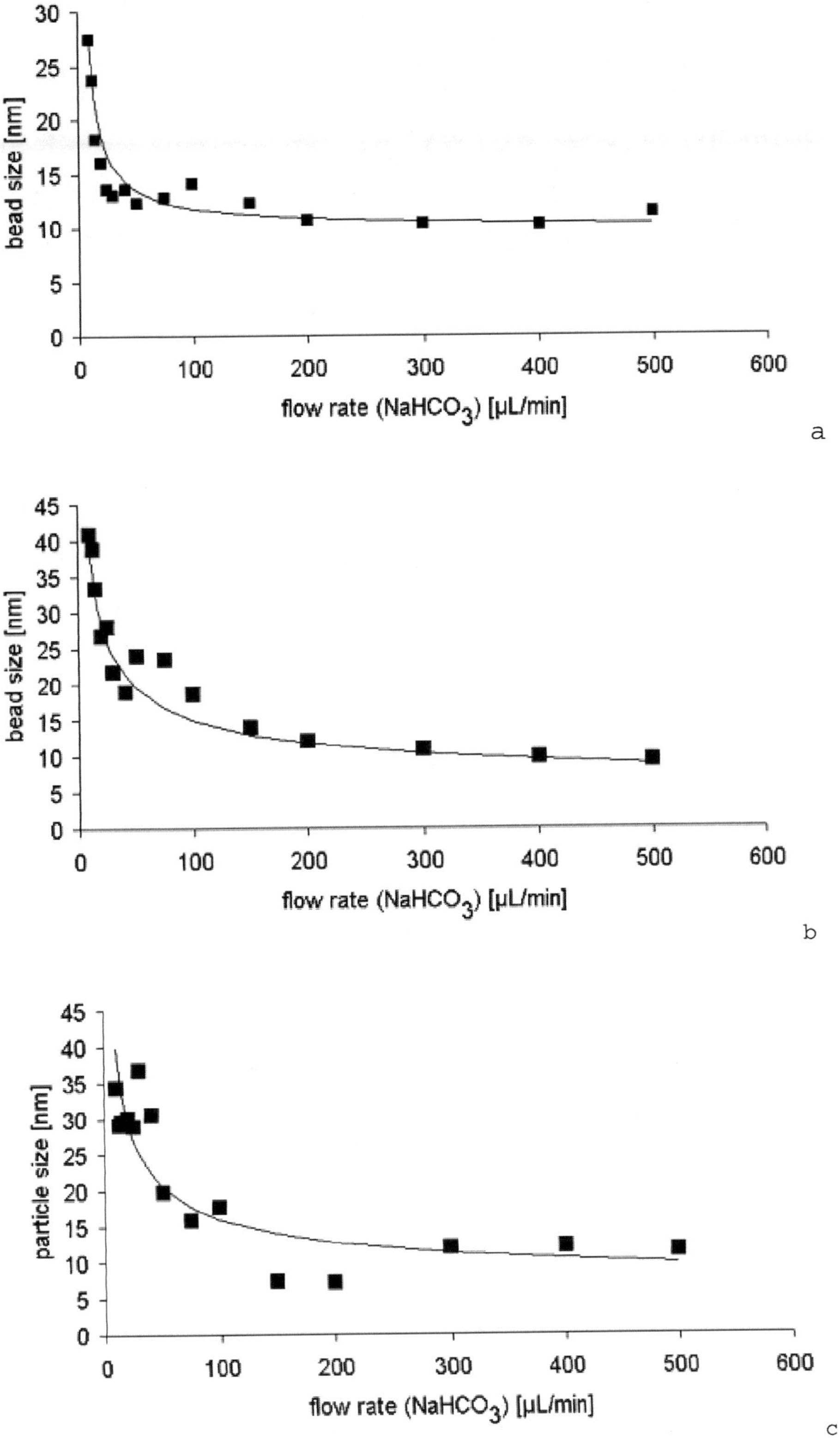

Fig. 4
Influence of pH on size of nanoparticles obtained by reduction of tetrachloroaurate with ascorbic acid at room temperature: a) flow rate of ascorbic acid (0.1 M): 25 µL/min, flow rate of HAuCl4 (1mM): 100 µL/min, b) flow rate of ascorbic acid (0.1 M): 50 µL/min, flow rate of HAuCl4 (1mM): 100 µL/min, c) flow rate of ascorbic acid (0.1 M): 25 µL/min, flow rate of HAuCl4 (1mM): 100 µL/min, addition of polyvinylpyrrolidone

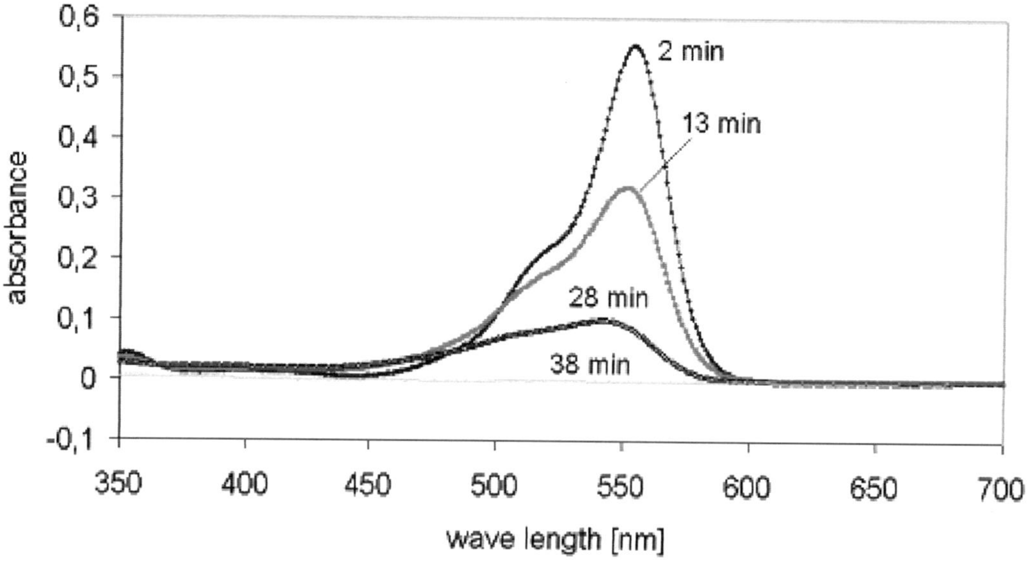

Fig. 5
Spectra of reaction mixture during the bleaching process of aqueous dye solution (rhodamine G) by oxidation with potassium peroxodisulphate in the presence of Au/Ag nanoparticles

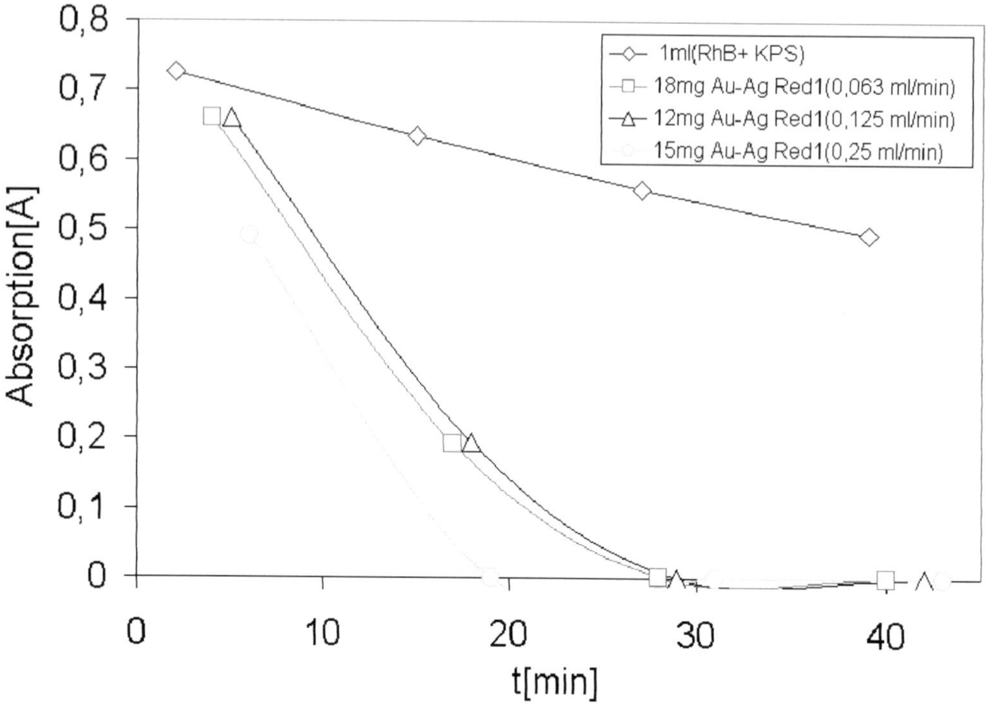

a

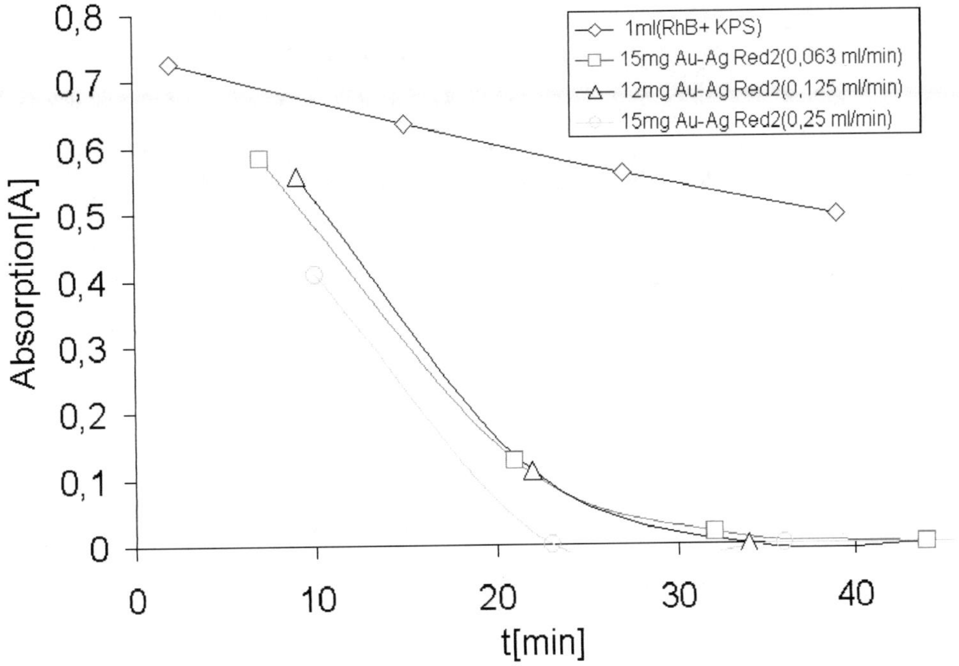

b

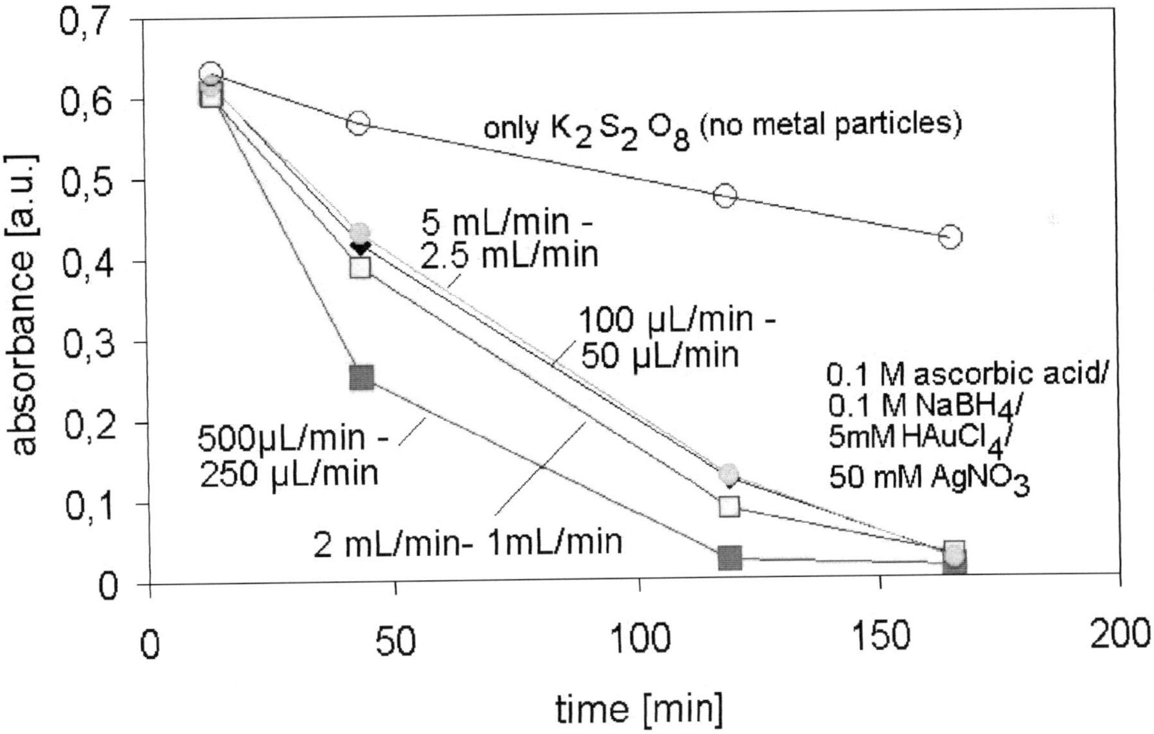

c

Fig. 6
Effect of Au/Ag nanoparticles on oxidative bleaching of dye solution: a) preparation of nanoparticles by borohydride and presence of SDS , b) preparation of nanoparticles by borohydride and presence of ascorbic acid, c) preparation of nanoparticles by reduction with ascorbic acid (Different flow rates and references without nanoparticles)

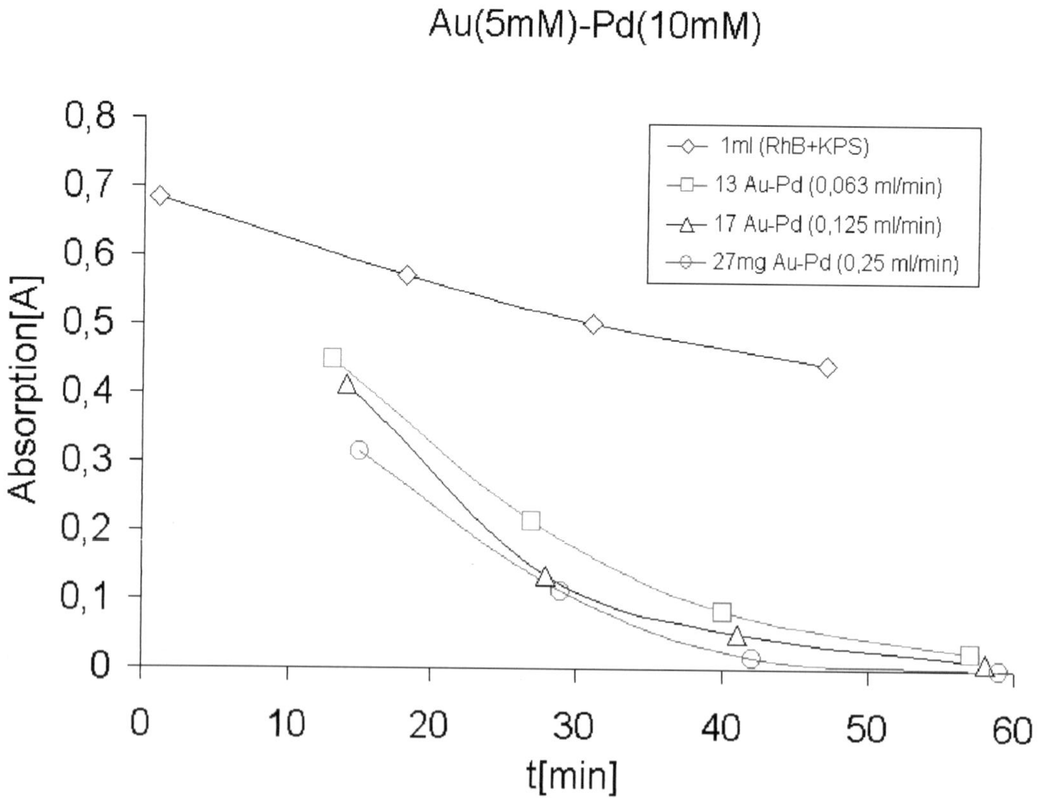

a

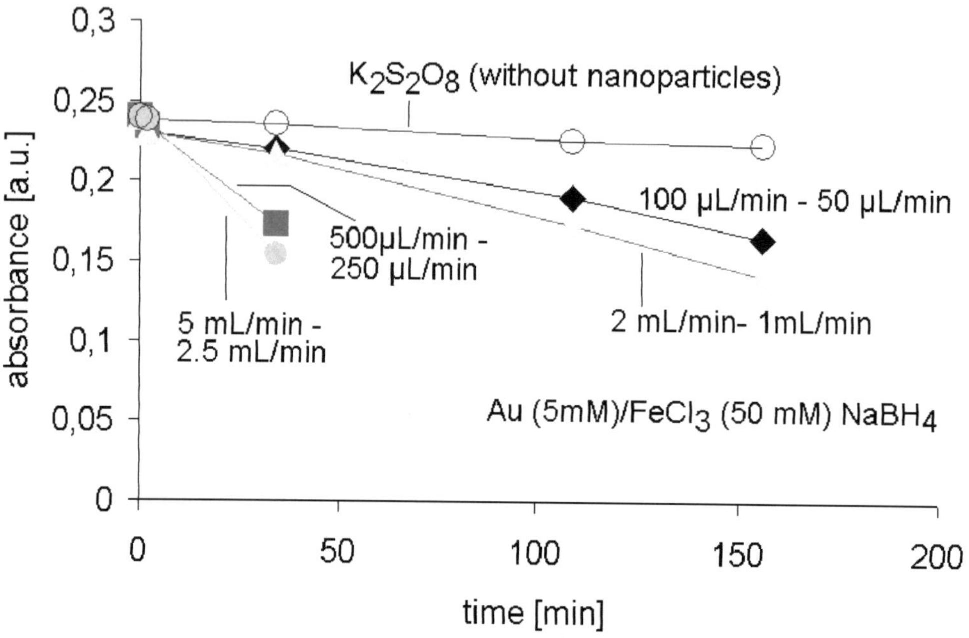

b

Fig. 7

Effect of Au/Pd and Au/Fe precipitates on oxidative bleaching of dye solution by potassiumperoxodisulphate: a) Au/Pd, b) Au/Fe (preparation at different flow rates and references without metal additives)

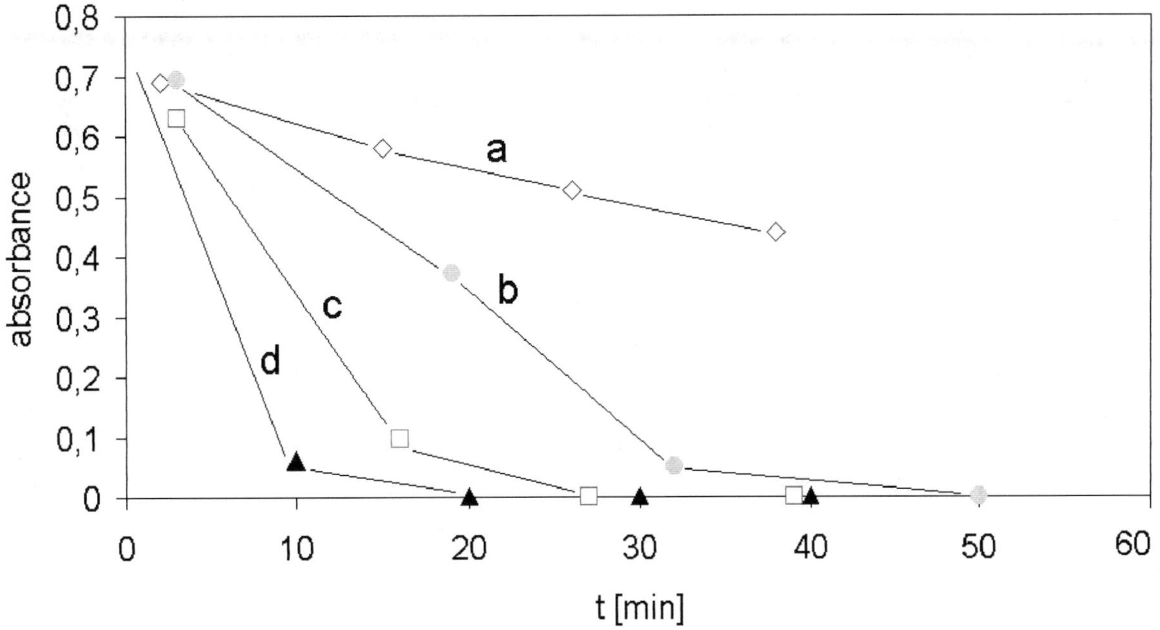

Fig. 8
Effect of reaction steps on bleaching activity of Au/Ag nanoparticles on rhodamine G in the presence of potassiumpersulphate (flow rate of reactants: 63 µL/min, flow rate of carrier (PP9): 125 µL/min): a) without metal nanoparticles, b) 1. reaction step: formation of gold nanoparticles (5mM $HAuCl_4$), 2. reaction step: deposition of silver (50 mM $AgNO_3$), c) 1. reaction step: formation of gold nanoparticles, 2. reaction step: deposition of silver and gold from a mixed solution (2.5 mM $HAuCl_4$, 25 mM $AgNO_3$), d) 1. reaction step: core formation from a mixed solution (2.5 mM HAuCl4, 25 mM $AgNO_3$), 2. reaction step: deposition of silver (50 mM $AgNO_3$)

Emulsification with Micro-Engineered Devices

Rob G. H. Lammertink, Maik J. Geerken, and Matthias Wessling. Science and Technology, University of Twente, Enschede, NL-7500AE, Netherlands

The preparation of emulsions with conventional technologies requires tremendous energy inputs of which the majority is lost in heat generation. This is a direct consequence of the high shear rates, necessary for droplet breakup. Recently, microstructured devices have gained interest for applications in emulsion fabrication. This originates from the length scale of microstructures being comparable to desired droplet sizes. As such, droplet formation is strongly controlled and requires less energy. By using several micro engineered devices, mono-disperse emulsions were fabricated. First, cross flow membrane emulsification with well defined membranes is described. This method relies on the shear induced droplet detachment from a pore through which the dispersed phase is fed. Membrane surfaces that are not wetted by the dispersed phase are needed for this. Besides, the drop formation at a given pore influences the cross flow profile locally and thus the drop formation at nearby pores. Even for very uniform pore size distributions, relatively polydisperse emulsions were obtained. This was demonstrated by using different membranes with different pore sizes and geometries. Secondly, we studied emulsification in a porous channel where the continuous phase is fed through the porous walls of the channel. The continuous phase breaks up a jet of liquid into discrete drops. This method yields, as the flow focusing method, monodisperse emulsions. The size of the droplets is mainly determined by the radius of the channel.

Membrane Micro Reactors for Heterogeneously Catalyzed Gas-Liquid Reactions

Jorrit de Jong, Rob G.H. Lammertink, and Matthias Wessling. Faculty of Science and Technology, University of Twente, P.O. BOX 217, Enschede, NL-7500AE, Netherlands

A current trend in the field of micro reactor engineering is the integration of multiple unit operations in a single device, such as pretreatment of raw materials, mixing, reaction and separation. We will present different micro reactors with integrated membrane functionality. These polymeric micro reactors are fabricated by CNC milling and clamping of commercial membranes. In this research project, we are mainly interested in applications based on gas-liquid contacting. Hydrophobic membranes are applied to create a stable interface, where the liquid cannot wet the porosity. By using multiple gas-liquid contacting points with different types of gasses or vapors, local differences in chemical composition can be realized in the micro reactor. The opportunities of this concept will be discussed into detail and we will give a proof-of-principle by generating local pH gradients. Then we will focus on heterogeneously catalyzed glucose oxidation. For this application, a Pt catalyst is immobilized on the membrane and oxygen is supplied from the gas phase. Analysis of the product stream is performed on-line by Electro Spray Ionization Mass Spectroscopy (ESI-MS). The tested micro reactor is shown in Figure 1. We will describe the fabrication procedure of the device and show first results, ending with a general outlook on the applicability of our concept.

Web Page: mtg.tnw.utwente.nl/research/micro/

Advanced catalytic microstructured reactor for continuous chemical synthesis integrated with a separation step

E.V. Rebrov, M.J.F. Warnier, O. Muraza, M.H.J.M. de Croon, V. Hessel, J.C. Schouten

Department of Chemical Engineering and Chemistry, Eindhoven University of Technology, P.O. Box 513, 5600 MB Eindhoven, the Netherlands
j.c.schouten@tue.nl; www.chem.tue.nl/scr

Microstructured reactors are nowadays regarded as a separate class of chemical reactors with their own specific characteristics. Microreactors have a channel diameter of 50 - 500 micron, which offers numerous advantages in processing of fine chemicals when compared with traditional batch-scale synthesis. The improved heat and mass transfer properties typical of microfluidic systems enable the use of more intensive reaction conditions that result in higher yields than those obtained with conventional size reactors. Residence time and heat management in microreactors can be exactly tuned to avoid secondary reactions. This leads to a higher selectivity to desired products, especially for a series of sequential reactions. Finally, microreactors offer the capability to carry out flexible on-site production of fine chemicals at the point of demand.

The synthesis of a large number of fine chemicals, particularly in the field of flavor and fragrance chemistry and pharmaceuticals, involves the selective hydrogenation of α,β-unsaturated carbonyl intermediates to the unsaturated alcohols as a critical step. At present, 2-propanol is used as a solvent in this type of hydrogenations. Owing to the safety problems inherent in the use of volatile organic materials as well as environmental concerns, it has become important to reduce the dependency on organic solvents in favor of a reaction medium that is "green" (low volatility, low toxicity, easy recyclability, and sufficient or superior abilities as a reaction medium compared with existing organic solvents). Fluorous (i.e. highly fluorinated saturated organic) solvents offer attractive features due to their excellent characteristics for dissolving diatomic gases. Although hydrogenation reactions can be carried out under liquid/liquid/gas conditions, from a reaction rate standpoint it is advantageous to operate under two phase liquid/gas conditions. The phase transition temperature is a function of the phase molar ratio. For this reason it is important to know at what temperatures various fluorous and non-fluorous solvents become miscible. For a number of binary and ternary solvent systems, a consolute temperature above which phase separation can not occur, whatever the composition, was determined. Unfortunately, binary systems of 2-propanol with common fluorous solvents have consolute temperatures above 383 K, which is too high for most selective hydrogenation reactions. On the contrary, ternary systems with 2-propanol have a wide range of phase transition temperatures, which depends strongly on composition. Figure 1 shows the consolute temperature in a system 2-propanol – hexafluoro-2-propanol – perfluorodecalin. The addition

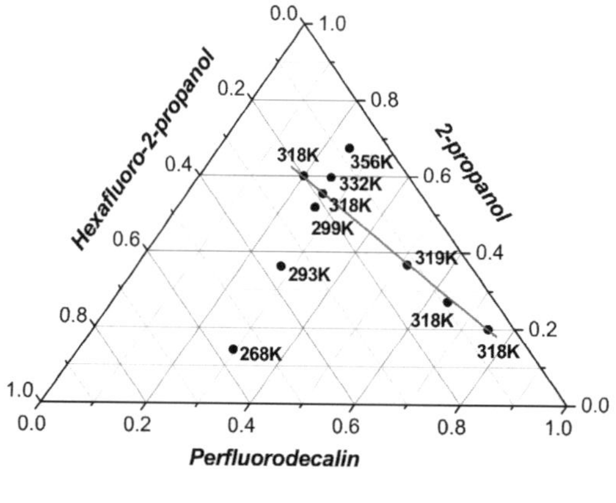

Figure 1. Phase transition temperature in the system 2-propanol – hexafluoro-2-propanol – perfluorodecalin (PFD). The red line corresponds to a consolute temperature of 318 K.

Table 1. Partition coefficients of major products of citral hydrogenation in a 2-propanol – hexafluoro-2-propanol – perfluorodecalin mixture as a function of temperature and molar composition. The partition coefficients are defined as ratio of substrate concentration in PFD to its concentration in a mixture of 2-propanol and HFR. Substrate concentration: 1 mol.%. Consolute temperature: 321 K.

	Composition: 2-propanol/HFP/PFD (molar)					
	27.2/9.0/63.8			55.0/18.4/26.6		
	296 K	308 K	315 K	296 K	308 K	315 K
citral	$2.2 \cdot 10^{-4}$	$4.0 \cdot 10^{-4}$	$3.5 \cdot 10^{-4}$	0.23	0.31	0.32
citronellol	$2.0 \cdot 10^{-6}$	$1.4 \cdot 10^{-4}$	$2.5 \cdot 10^{-4}$	0.22	0.28	0.29
citronellal	$6.7 \cdot 10^{-4}$	$1.5 \cdot 10^{-3}$	$1.4 \cdot 10^{-3}$	0.23	0.32	0.33
nerol	$2.0 \cdot 10^{-6}$	$2.7 \cdot 10^{-4}$	$1.2 \cdot 10^{-4}$	0.22	0.28	0.29
geraniol	$2.0 \cdot 10^{-6}$	$4.0 \cdot 10^{-5}$	$2.5 \cdot 10^{-4}$	0.22	0.28	0.29

of an organic substrate slightly increases the phase transition temperature: e.g. an addition of 1 mol.% of citral to a system corresponding to the red line in Figure 1 increased the consolute temperature by 3 K from 318 to 321 K. In order to extract the products from reactions involving fluorous solvents in a rational way, partition coefficients between the fluorous and organic phases were determined (Table 1). It can be seen that a molar composition of 2-propanol/HFP/PFD of 27.2/9.0/63.8 allows effectively extracting the major reaction products of citral hydrogenation to a 2-propanol – HFP mixture after cooling to 296 K.

The product and unreacted reactant dissolved in an organic solvent were separated from hydrogen using a passive gas-liquid microstructured separator positioned downstream of the microreactor (Figure 2). The separator chip is fabricated in a 500 micron thick glass wafer and a 100 micron thick membrane with 12 micron wide holes for the gas release was fabricated in silicon. After teflonizing with a hydrophobic polymer and subsequent Piranha cleaning, the silicon wafer was anodically bonded to the glass wafer. The difference between the system and the hydrogen disposal side was kept below 0.025 bar at 298 K by controlling the flow rate. The organic solvent can flow through the hydrophilic membrane easily because there is no gas-liquid interface (Figure 2A). The fluorous solvent can flow through the hydrophobic membrane (Figure 2B). Because the system is slightly pressurized, the edge of the gas bubbles deforms so that it conforms to the contour of the opening. The thin liquid film breaks up and creates a passage for gas discharge (Figure 2C). Since one side of the gas separator is hydrophobic, the hydrophilic liquid forms a meniscus at the inner mouth of the opening and cannot flow into the hole (Figure 2D).

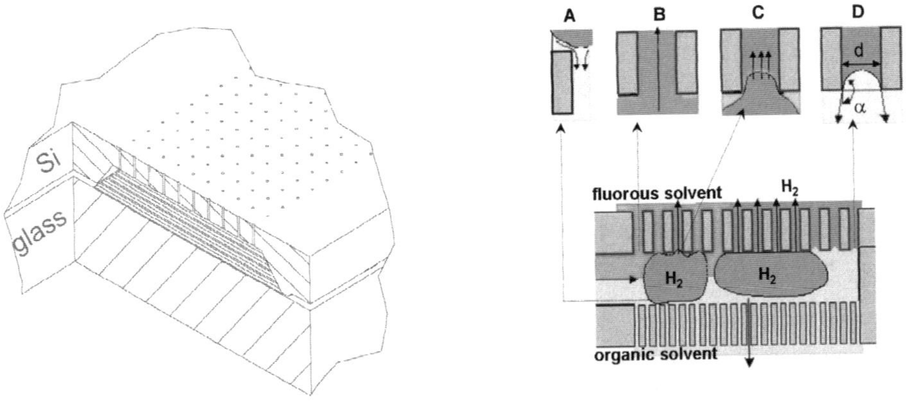

Figure 2. Schematic view of a gas-liqiud microseparator and the gas-liquid separation mechanism. The openings in the top Si membrane are 12 micron in diameter.

With the implementation of a separation step, the whole microstructured device consists of three individual modules which are connected to each other (Figure 3). The mixer module is used to warm up a gas-liquid mixture to the reaction temperature. Then, a single liquid phase with high hydrogen solubility enters the reactor module where a hydrogenation reaction take places on a catalytic coating deposited on the channel walls. A separator module is positioned downstream of the reactor module.

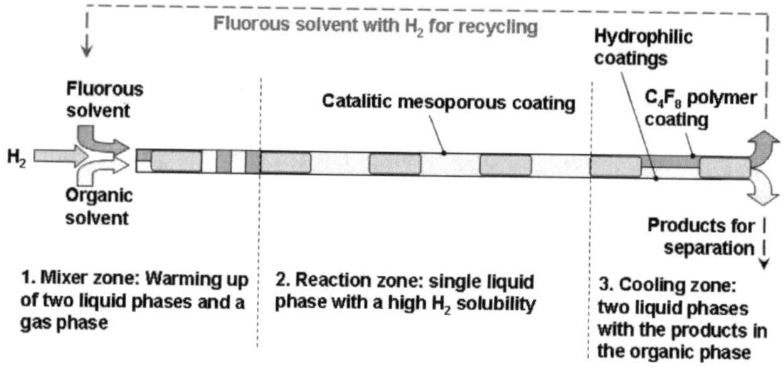

Figure 3. *Schematic view of a modular design of micromixer, microreactor and microseparation units.*

To prevent the G/L separator chip from high pressure at the exit of the microreactor, a fast safety valve (release time: 5 ms) and a 2 m safety capillary is inserted between the microreactor and the microseparator (Figure 4). If the pressure at the reactor outlet exceeds the critical pressure of 2 bar, the safety valve is opened and the pressure is released to the environment. The safety capillary is required to provide enough time for the valve to be opened and to prevent that a high pressure peak reaches the microseparator. This time depends strongly on the flow pattern in the safety capillary. Usually, the Taylor flow regime is realized. In this regime, the pressure drop across the safety capillary is a function of both slug length and the gas hold-up. In the paper, the results with respect to different modules of the microreactor/microseparator system will be presented.

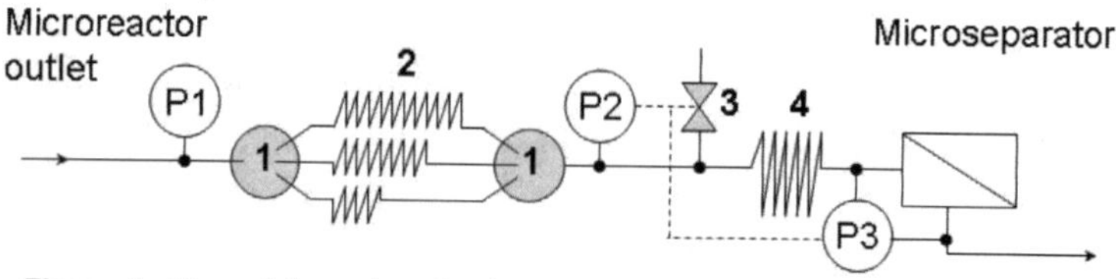

Figure 4. *The safety system for the G-L microseparator: (1) rotary multiposition valve for selectively directing the flow through a plurality of paths; (2) pressure reducing grid: depending on the outlet pressure in the microreactor, a capillary providing the desired pressure drop is chosen; (3) fast pressure release valve with an electrical actuator; (4) safety capillary; P1, P2 – pressure sensors, P3-differential pressure sensor.*

References

[1] V. Haverkamp, V. Hessel, H. Löwe, G. Menges, M.J.F. Warnier, E.V. Rebrov, M.H.J.M. de Croon, J.C. Schouten, M. Liauw, Chem. Eng. Technol., 29 (2006) 1015-1026.

[2] M.J.F. Warnier, E.V. Rebrov, M.H.J.M. de Croon, V. Hessel, J.C. Schouten, *Gas hold-up in Taylor flow in rectangular micro channels*, Chem. Eng. J., submitted.

Olefins by High-Intensity Oxidation in Microchannel Reactors

Terry Mazanec[1], Tom Yuschak[1], Bin Yang[1], Richard Long[1], and Anna Lee Tonkovich[2]. (1) Velocys, Inc., 7950 Corporate Boulevard, Plain City, OH 43064, (2) Technology Development, Velocys, Inc., 7950 Corporate Boulevard, Plain City, OH 43064

Ethylene is the backbone of the chemical industry, both literally and figuratively; over 120 million tons are produced annually with a value of over $100 billion, and ethylene is the basic building block of many polymers and commodity chemicals. Ethylene is the simplest olefin, a family of chemical intermediates used to produce a wide variety of plastics and end use chemicals. The conventional path to ethylene is via steam cracking, which involves heating ethane and steam in 60 foot long 4 inch diameter tubes hung in a cathedral-like furnace to form ethylene, propylene, and by-products. Despite years of incremental improvements, this approach is inefficient, and suffers from poor selectivity to ethylene and other valuable products. Velocys, Dow Chemical, Pacific Northwest National Laboratory (PNNL), and U.S. Department of Energy (DOE) are developing a process for the production of olefins via oxidative dehydrogenation in highly heat-integrated microchannels. This approach offers improved feedstock utilization and substantial energy savings due to microchannel architecture's unique ability to control reaction temperatures. Since ethylene is such a valuable commodity, it has been the focus of intense research and development. However, all attempts to improve the process, including many attempts at oxidative dehydrogenation, have failed to meet economic targets. The challenge facing all attempts to oxidize paraffins to ethylene is achieving high product yields, while overcoming the cost associated with supplying pure oxygen. At project initiation, scientists at Dow recognized that an oxidative dehydrogenation process could be commercially attractive if an improved product mix could be obtained. This has been achieved by exploiting the time-temperature-oxygen content trajectory possible only in microchannel architecture. Within a microchannel one can more effectively manage reaction time, temperature and the introduction of oxygen, by carefully controlling flow rates and reactor geometry, thus defining the time-temperature-oxygen content trajectory of the process. Breakthrough performance has been achieved with a reactor design, which was accelerated by modeling at three scales. First, a simplified kinetic model was developed from a complex reaction network of over 100 elementary steps by selecting a manageable few that account for key products and most of the energy generation and consumption. Second, the reduced network of reactions was used in a model of a slice of a commercial reactor that includes reactor hydrodynamics with detailed local mass and heat balances. Third, a global model was used to aggregate performance of many parallel reactor slices, as predicted by the intermediate model, to assess overall performance of a commercial full-scale reactor. This multi-scale modeling methodology closely predicted the experimental results of an integrated microstructured reactor with 84% ethane conversion and greater than 75% ethylene selectivity. The goal of the project is to demonstrate the economic viability of this approach. And this goal has been achieved. Although much work remains, laboratory experiments have shown that economic targets can be met in microchannel hardware with specially adapted catalysts. This presentation will describe the achievements of this DOE funded project and its implications for the broad applicability of microchannel process technology to selective oxidation processes.

Kinetics of Hydrogen Peroxide Synthesis in a Microreactor by Direct Combination of Oxygen and Hydrogen

Yury Voloshin[1], Adeniyi Lawal[2], and Raghunath Halder[2]. (1) Department of Chemical, Biomedical and Materials Engineering, Stevens Institute of Technology, New Jersey Center for Microchemical Systems, 1 Castle Point, Hoboken, NJ 07030, (2) New Jersey Center for Microchemical Systems, Dept. of Chemical, Biomedical and Materials engineering, Stevens Institute of Technology, Castle Point, Hoboken, NJ 07030

Hydrogen peroxide was formed in a packed-bed microreactor in a three-phase process involving the direct combination (DC) of hydrogen and oxygen over a solid catalyst in the presence of a liquid solvent. Hydrogen and air were combined in all proportions, including explosive regime, over a platinum group catalyst supported on an oxide. The DC process comprises multiple reactions involving formation of hydrogen peroxide, parallel formation of water, hydrogen peroxide reduction by hydrogen, and decomposition by disproportionation. Synthesis of hydrogen peroxide was conducted at conditions where the selectivity for H_2O_2 was 100%, so that all reactions except for H_2O_2 synthesis could be neglected. A Langmuir-Hinshelwood rate expression for the synthesis of hydrogen peroxide was determined as part of a program to determine the overall kinetics of hydrogen peroxide formation.

Production of H_2O_2 by Controlled H_2/O_2 Reaction in a Microchannel Reactor

Adeniyi Lawal[1], Yury Voloshin[1], and E. Dada[2]. (1) New Jersey Center for Microchemical Systems, Dept. of Chemical, Biomedical and Materials engineering, Stevens Institute of Technology, Castle Point, Hoboken, NJ 07030, (2) FMC Corporation, US 1 & Plainsboro Road, Princeton, NJ 08543

Hydrogen peroxide was formed in a microreactor in a three-phase process involving the direct combination (DC) of hydrogen and oxygen over a solid catalyst in the presence of a liquid solvent. Hydrogen and air were combined in all proportions, including explosive regime, over an in-house formulated platinum group catalyst supported on an oxide. In addition to packing the microchannel with the catalyst in particulate form, we developed closed-channel flow coating, open-channel surface-selective and dip coating methods to deposit sol-gel-based thin-film catalysts into microchannels, and cellular structures. H_2O_2 has been produced in concentrations of significant commercial interest at moderate pressure and temperature conditions and a residence time that is almost two orders of magnitude less than that for macroreactors. The multiple reactions involved in the DC process have been identified, and the reaction mechanisms elucidated with the goal of determining the overall kinetics of the hydrogen peroxide formation. Based on the extensive data acquired from the laboratory studies, an optimized process analysis, design and simulation was developed for a pilot-plant suitable for on-site production of H_2O_2 at production rates of commercial interest. The pilot-plant including the multichannel, multilayer reactor system for the demonstration of the feasibility of on-site DC process for H_2O_2 was constructed, and preliminary data obtained. This presentation will discuss the accomplishments as well as the challenges of the project.

The Use of Immobilised Catalysts for Organic Synthesis in Flow Reactors

P. Watts, Charlotte Wiles, and Linda Woodcock. University of Hull, Hull, United Kingdom

Synthetic chemists are continually searching for new ideas that make the construction of organic compounds easier. While solid-supported reagents, scavengers and catalysts have been used to great effect speeding up the purification of product mixtures, reactions are generally slower and employ laborious filtration techniques to remove the supported reagent post reaction. Furthermore, literature president dictates that the efficiency of supported reagents deteriorates over time as a result of mechanical degradation. In comparison, supported reagents are ideally suited for use in flow reactors where compounds are conveniently synthesized by the flow of substrates over the immobilized reagent or catalyst. In such systems, the rate of reaction is enhanced as a result of the high surface to volume ratio and as the catalyst is held within the reactor, it is not necessary to filter the reaction products to remove the supported reagent, enabling several of the traditional disadvantages to be overcome with ease.

Very few publications have reported organic synthesis in micro reactors incorporating supported reagents or catalysts; this is attributed largely to the difficulties associated with the fabrication of such devices. The Hull group has recently overcome these problems using a variety of approaches; the most successful being the incorporation of supported reagents between porous silica frits after the micro reactor has been fabricated. In such systems the group has demonstrated that a wide variety of reaction types may be conducted in near quantitative yield obtaining products in high purity, straight from the reactor, without the need for any form of purification. As the immobilized reagent suffers less physical damage in flow reactors compared with conventional stirred reactions, the result being that the reagents last longer and reactions are more reproducible.

The Hull group has also investigated the feasibility of incorporating supported reagents in series within continuous flow reactors in order to perform multi-step syntheses. In this presentation a series of other reactions will be presented including the use of immobilised enzymes in such systems.

Enzyme Immobilization in Chemical Microsystems at Louisiana Tech

James Palmer[1], Shaila Reddy[1], and L. Dale Snow[2]. (1) Chemical Engineering, Louisiana Tech University, 600 W Arizaon, Ruston, LA 71272, (2) Chemistry, Louisiana Tech University, 600 W Arizaon, Ruston, LA 71272

The use of microfluidics for studying enzyme immobilization using layer-by-layer nano self-assembly will be presented. The highly controlled liquid/solid interface due to the laminar regieme present in microfluidics is ideal for these studies. Urease has been immobilized on silicon microchannel walls to catalyze the reaction of urea to ammonia. The Berthelot procedure is used to directly measure the ammonium ion concentration of the effluent to determine product conversion. Enzyme activity and stability is very sensitive to the choice of polyelectrolytes used for immobilization and will be discussed. Results of Quartz Crystal Microbalance studies and protein assay measurements will also be presented.

Methanol Oxidation Mechanism by Ion-Modified Methanol Dehydrogenase Enzymes for Fuel Cell Applications

Nagesh B. Idupulapati and Daniela S. Mainardi. Louisiana Tech University, 911 Hergot St., Ruston, LA 71272

Bacterial Methanol Dehydrogenase (MDH) is a water-soluble quinoprotein that oxidizes methanol and other primary alcohols to their corresponding aldehydes. The crystal structure of MDH from Methlylobacterium extorquens [1, 2] and from Methylophilus W3A1 [3-5] has been characterized and it has been determined that the enzyme active center contains a Ca2+ ion, pyrroloquinoline quinone (PQQ) and amino acids. It has been suggested that apart from holding the PQQ molecule in place in the active site, the calcium ion might have an important role in the methanol oxidation mechanism. Some authors [6] have used Ca2+-free MDH enzymes to obtain enzymes containing Sr2+ and Ba2+ in their active sites. Their experimental results have shown that there are no major differences between these enzymes in the interactions between PQQ and the metal ions in the active site. However, even though the Ba2+-modified enzyme has a relative low affinity for methanol, its activation energy for the oxidation reaction is half that of the normal Ca2+-containing MDH enzyme.[6] This result was not expected since the replacement with Ba2+, a weaker Lewis acid, should decrease the activity of the enzyme, and therefore increase its activation energy.[6] This means that Ba2+ has the potential to activate PQQ, but apparently subtle differences in the active site of PQQ-containing enzymes determine the way and the extent to which the activation is expressed. Fuel cells that use bacterial MDH enzyme as anode catalyst are potential attractive power sources, which may produce continuous power output of about 100 mW. Understanding the role of the ion in the active site of MDH, as well as the fuel oxidation mechanism will help to contribute to the design of novel catalysts for fuel cell applications. In this work, the methanol oxidation mechanisms by Ca2+- and Ba2+-MDH enzymes are investigated using Quantum Mechanical and Molecular simulations. Quantum mechanical Density Functional Theory (DFT) simulations are used to obtain geometry, energetic, electronic configurations, and binding energies of small portions of the active site of the enzyme. Equivalent information is obtained on larger systems using Ab Initio Molecular Dynamics. The activation energies for the methanol oxidation reactions are calculated and rates predicted using Transition State Theory. Our findings suggest that the nature of the ion modifies the binding and orientation of methanol with respect to the active site of the enzyme, facilitating the methanol oxidation reaction in the case of having Ba2+ ion in the active site of MDH enzyme.

1. Ghosh, M., C. Anthony, K. Harlas, M.G. Goodwin, and C.C.F. Blake, The Refined Structure of the Quinoprotein Methanol Dehydrogenase from Methylobacterium Extorquens at 1.94 Å. Structure (London), 1995. 3: p. 1771-1787. 2. Afolabi, P.R., M. F., K. Amaratunga, O. Majekodunmi, S.L. Dales, R. Gill, D. Thompson, J.B. Cooper, S.P. Wood, P.M. Goodwin, and C. Anthony, Site-Directed Mutagenesis and X-Ray Crystallography of the PQQ-Containing Quinoprotein Methanol Dehydrogenase and Its Electron Acceptor, Cytochrome CL. Biochem., 2001. 40: p. 9799-9809. 3. Xia, Z. X., Y.N. He, W.W. Dai, S. White, G. Boyd, and F.S. Mathews, Detailed Active Site Configuration of a New Crystal Form of Methanol Dehydrogenase from Methylophilus W3A1 at 1.9 Å Resolution.

Biochem., 1999. 38: p. 1214-1220. 4. White, S., G. Boyd, F.S. Mathews, Z.X. Xia, W.W. Dai, Y.S. Zhang, and V.L. Davidson, The Active Site Structure of Calcium Containing Methanol Dehydrogenase. Biochemistry, 1993. 32: p. 12955-12958. 5. Xia, Z.X., W.W. Dai, Y.S. Zhang, S. White, G. Boyd, and F. S. Mathews, Determination of the Gene Sequence and the Three-dimensional Structure at 2.4 Å Resolution of Methanol Dehydrogenase fromMethylophilusW3A1. J. Mol. Biol., 1996. 259: p. 480-501. 6. Goodwin, M.G. and C. Anthony, Characterization of a novel methanol dehydrogenase containing a Ba2+ ion at the active site. Biochem J., 1996. 318(2): p. 673-679.

Synthesis and Assembly of Size-Controlled Polymer Beads and Capsules in an Axisymmetric Co-Flowing Microsystem

Christophe Serra[1], Michel Bouquey[1], Laurent Prat[2], and Georges Hadziioannou[1]. (1) Ecole de Chimie Polymères et Matériaux - Laboratoire d'Ingénierie des Polymères pour les Hautes Technologies, ULP - CNRS UMR 7165, 25, Rue Becquerel, Strasbourg Cedex-2, F-67087, France, (2) Laboratoire de Génie Chimique, CNRS UMR 5503 - INPT, 5 rue Paulin, Toulouse, 31106, France

We will present an axisymmetric microreactor for the synthesis of monodisperse polymer particles (CV<5%), either beads or capsules, in the size range of tens to several hundred of microns. The emulsification of the monomer phase is achieved in a very simple needle/tubing co-flowing microsystem without any surfactant or surface treatment. Polymer beads are obtained after UV irradiation while polymer capsules are directly generated at the needle tip through an interfacial polycondensation reaction. A preliminary study has shown that in the dripping regime the variation of particle diameter with respect to operating parameters (flow rates and viscosities of the dispersed and continuous phases) is fairly well described by an empirical relationship between two dimensionless numbers, the normalized particle diameter and the Capillary numbers ratio. This co-flowing microsystem also allows for the preparation of inorganic-organic hybrid beads as well as hollow and doped capsules. By using two co-flowing microsystems in parallel, two different types of particles can be obtained and assembled downstream to form polymer particles necklaces with different arrangement, alternate, block and random.

Synthesis and Production of Polyimide Fine Beads by the Use of Micro-flow-mixing Technique

Katsuhiko Tahara[1], Jiro Mizuya[1] and Masaaki Sato[2]

[1] *Research Center, Arakawa Chemical Industries, Co. Ltd., 1-1-9, Tsurumi, Tsurumi, Osaka 538-0053, Japan*

[2] *Department of Chemistry, Faculty of Liberal Arts and Sciences, Osaka Prefecture University, Nakaku Gakuencho 1-1, Sakai, Osaka 599-8531, Japan*

Prepared for presentation at AIChE Spring National Meeting/ April 26, 2007/ Topical 1: Applications of Micro-reactor Engineering/ Particles and Beads

Unpublished

Abstract

Fine polyimide beads were obtained by thermal dehydration reaction of fine poly(amic acid) beads, which were prepared by mixing tetracarboxylic acid dianhydride solution with diamine solution using micro-flow-mixing devices. The diameter of these beads, which could be controlled from submicron to several microns, had a coefficient of variation less than 10%.

1. Introduction

Polymer beads are now widely used for materials of various kinds of chemical industry products, such as cosmetics, paints, color inks, binder, electronic display, and medical supplies. The performance of these products is known to be affected by the particle properties, such as composition, size, size distribution, and shape. Polymer beads made of polystyrene, poly(methyl metacrylate), poly(acrylic acid), polyurethane, polyimide are widely known. In the case of polyimide particle, its heat-resistance property is of special interest.

In recent years, microreactor is gaining increasing interest since its efficient properties are expected to open a new field in chemical production. As the microreactor has advantages of high surface to volume ratio, excellent heat and mass transfers, and improved safety due to its small quantity, it can realize rapid mixing at a precise temperature.

Here we wish to report a new method for obtaining mono-disperse like spherical shaped polyimide fine beads by the use of the microreactor.

2. Experimental

The reaction scheme to get polyimide through poly(amic acid) is shown in Eq. 1. Fig. 1 shows the overview of the experimental setup for preparing the poly(amic acid) beads.

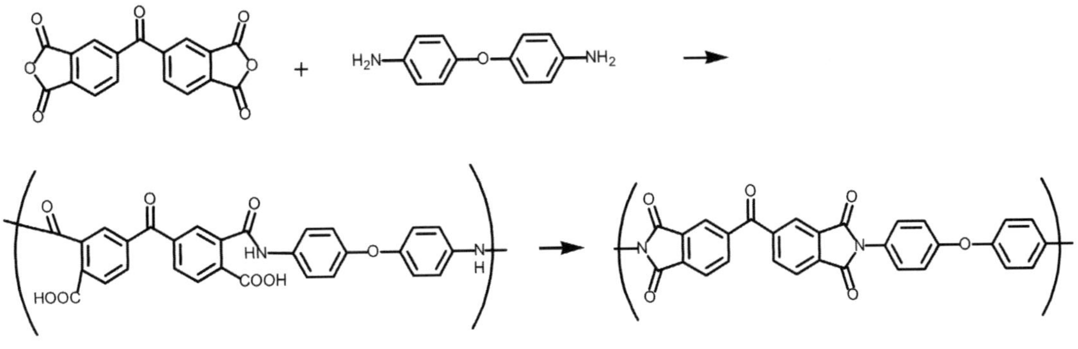

Eq. 1 Synthetic pathway to prepare poly(amic acid) and polyimide.

An acetone solution of benzophenone-3,3',4,4'-tetracarboxylic dianhydride (BTDA, 0.04 mol/L) was mixed with another acetone solution of 4,4'-diaminodiphenyl ether (DPE, 0.04 mol/L) at the microreactor by the use of a syringe pump. The mixed solution was collected in a tank, where the beads formation went essentially to completion in about 10 minutes. The beads of poly(amic acid) were isolated by filtration. Then, the beads were washed with acetone followed by drying at room temperature *in vacuo*. The size and shape of the beads were analyzed by Scanning Electron Microscope (SEM) as shown in Fig. 2.

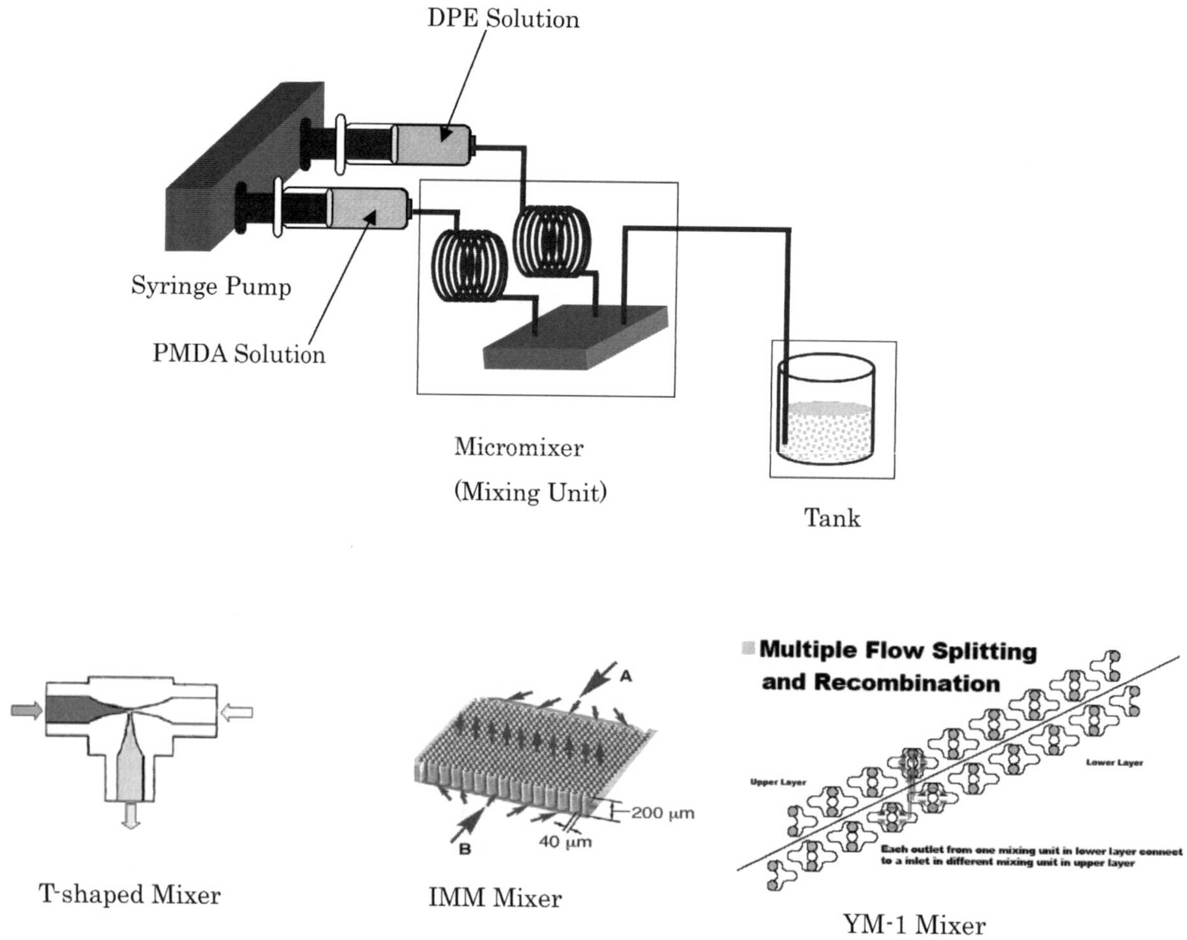

Fig. 1 Experimental setup and three types of microreactors

3. Results and Discussion

First, we studied on the mixing procedure. When a conventional mixing apparatus, a flask equipped with a dropping funnel, was used in the synthesis of poly(amic acid), only amorphous precipitates were obtained. Mixing the BTDA and DPE solutions with the use of a micro-flow technique, the shape of the resulting poly(amic acid) was greatly improved. SEM images of the poly(amic acid) were shown in Fig. 2. The micro-flow-mixing with a T-shaped mixer gave rise to semi-spherical shaped beads as shown in Fig. 2(a). When the mixing was performed with an IMM mixer or a YM-1 mixer, fine spherical shaped beads of poly(amic acid) were obtained as shown in Fig. 2(b) and (c), respectively. Table 1 shows qualitative analysis concerning median size and coefficient of variation (CV). In the case of the micro-flow system with the IMM mixer or YM-1 mixer, very fine poly(amic acid) beads, median size of 0.79-0.97 μm and CV less than 10 %, were obtained in a 60-70 % yield.

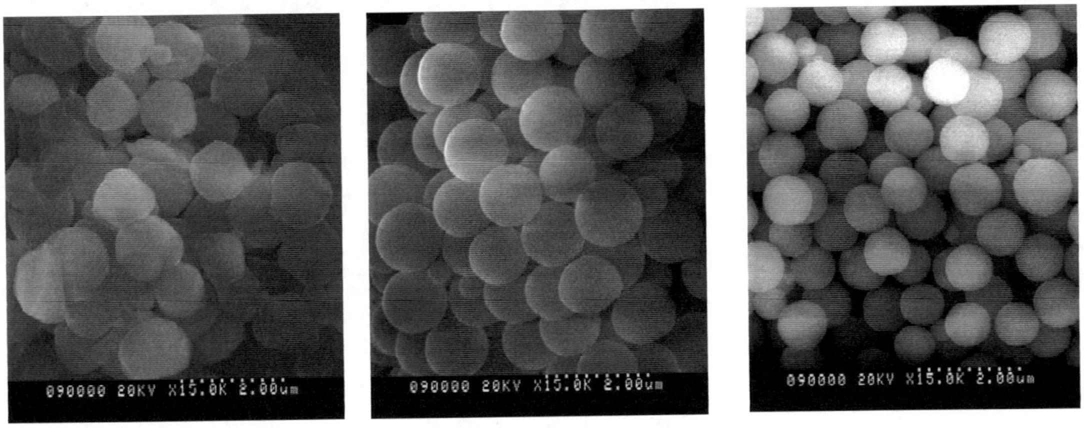

(a) T-shape mixer (b) IMM mixer (c) YM-1 mixer

Fig. 2 SEM images of three poly(amic acid) beads prepared by different micromixers

Table 1 Qualitative analysis of poly(amic acid) beads.

Mixing		Median Size (μm)	CV (%)
Micro-flow	T-shaped	1.07	14.80
Micro-flow	IMM	0.97	7.10
Micro-flow	YM-1	0.79	5.95
Dropping funnel		1.31	14.60

The mixing was performed at 293K.

To convert the poly(amic acid) into polyimide, the beads thus obtained were heated up to 473 K *in vacuo* for 6 hours. The dehydration reaction to give polyimide were identified by IR measurement. During the dehydration reaction, the median size decreased about

10 %. The resulting polyimide beads maintained the fine spherical shape as well, which was checked by SEM analysis.

Table 2 shows effects of triethylamine (NEt_3) additive at 323 K. When NEt_3 was added as a base, the nucleophilicity of the amino group of DPE was kept intact during the polymer formation by suppressing protonation to the amino group. The NEt_3 additive had large effects on median size, CV, and yield as shown in Table 2. By adding the NEt_3, median size became smaller with dramatic decrease in CV. The yield of the poly(amic acid), however, increased up to 81 %.

Table 2 Effect of NEt_3 additive

NEt_3 (mol %)	Yield (%)	Median Size (μm)	CV (%)
0	67	2.43	17.07
10	75	1.44	6.73
50	81	1.42	5.78

The mixing was performed with the IMM mixer at 323 K.

4. Conclusion

Fine spherical polyimide beads were obtained by thermal dehydration reaction of fine spherical poly(amic acid) beads, which were prepared by mixing tetracarboxylic acid dianhydride solution with diamine solution using micro-flow-mixing devices, IMM mixer and YM-1 mixer. The diameter of the beads ranging from submicron to several microns could be controlled by applying suitable reaction conditions. The CV of less than 10 % implied a narrow distribution of the bead size. Addition of the NEt_3 base was thought to bring about uniform polymerization by accepting a proton from a carboxylic acid of the poly(amic acid). Otherwise, the DPE was protonated to inactivate the nucleophilic polymerization. As a result, uniform polymerization took place and mono-disperse like spherical shaped beads were obtained.

Fluorescence-labelled polymer beads for chemical monitoring of nanoliter fluid segments

Anette Funfak [1)], Martin Fischlechner [2)], Edwin Donath [2)] and J. Michael Köhler [1)]
1)Technical University of Ilmenau, Institute of Micro and Nanotechnologies/Institute for Physics
and
2) University of Leipzig, Biophysical Department

correspondence:
Prof. Dr. M.Köhler, Techn. Univ., D-98693- Ilmenau, Weimarer Str. 32, D-98693 Ilmenau, Inst. f. Micro and Nanotechnologie, Inst f. Physics, Germany, phone 49-3677-693700, michael.koehler@tu-ilmenau.de,

Abstract

The introduction of dye-labelled and polyelectrolyte-coated polymer particles in nanoliter fluid segments is investigated. It could be shown, that particles of 3 µm up to 20 µm diameter can be brought into the fluid segments. Dye labelled-particles are suited for the labelling of segments in order to support the identification of a certain segment or for reconstructing the process history of a segment after a group of process steps. A high number of different segments can be distinguished by combinatorial colour codes. Polyelectrolyte-coated polymer particles can be used for the indication of chemical composition of fluid segments, if they contain indicator dyes in the surface layer. The immobilization of dyes in the ultrathin polyelectrolyte surface layer leads to very short diffusion paths inside the indicator material and results in short response times. The beads are usable for the recognition of chemical changes inside segments, for example for pH-shifts due to mixing, chemical reactions or metabolic activity of segment-incorporated biological objects like bacteria, cells and developing embryos.

Key words

fluid segments, labelling, chemical recognition, poylmer particles, sensor beads

1. Introduction

The application of liquid/liquid two-phase systems in micro fluidics attracts increasing interest due to the unique possibilities of manipulation of small process volumes. The single reaction compartments can be manipulated individually without cross-talk between them. Fluidic dispersion is completely suppressed and the fast formation of segments can easily be performed [1-3]. So, this technique is well suited for serial miniaturized processing. In consequence, nanoliter fluid segments are under investigation for combinatorial chemical synthesis [4], for analytical applications [5,6] and for miniaturized screening experiments [7,8]. Therefore, the precise definition and the measurement of concentration of reactants, dissolved catalysts or of effectors on biological processes is important. Particular, the cell activity, metabolism, cell division, growth, differentiation processes, embryonic development and cell-cell-communication are strongly effected by chemical conditions.

The application of classical analytical tools is difficult for fluid segments due to their small size. Despite a certain applicability of miniaturized sensors, there is a challenge for alternative read-out of chemical information by non-invasive techniques. It was known, that fluorescence dye-containing micro particles can be embedded in polymer films and applied as miniaturized sensors [9-11]. Sensor beads with hydrophilic surface layers are of particular interest for biological processes, biochemical operations and for chemical reactions in aqueous systems. Polymer particles of a non-hydrophilic core material coated by hydrophilic surface layers containing the indicator dyes are particular attractive due to there low swelling behaviour and short diffusion path ways for analyt from the solution to the immobilized indicator molecules in the particle shell [12-14]. Therefore, here the introduction of core-shell micro particles carrying indicator molecules in to fluid segments is reported.

2. Experimental

The experiments were carried out in microfluidic arrangements. A syringe pump actuation was applied for the generation of fluid segments as well as for the introduction of particles. The particles ("beads") can either be dispersed inside the segment liquid before the segmentation process or they can be introduced by a dosing process into the preformed segments through a micro injector. The bead-containing fluid segments were conducted through glass micro channel chips (IPHT Jena) for optical imaging and measurement of integral fluorescence intensity (fig.1).

The polymer matrix of the core of the particles consists on a hydrophobic polymer like polystyrene. The core material is inert and does not swell inside an aqueous environment. The particles can be composed of different materials varying in size and in composition in order to adapt approximately the density of the process liquids. The application of density-adapted particles is preferred for avoiding undesired sedimentation. The surface of the polymer core is covered by a polyelectrolyte shell. The charging of polymer films support the binding of ionic dyes by electrostatic forces (fig 2). The formation of multishell layers at the particle surfaces is advantageously in order to enhance the area density of immobilized indicator dye molecules. The multishell particles are formed by alternating deposition of cationic and anionic polyelectrolyte molecules. The character of last deposited polymer molecules determines the total charge of the whole polymer particle. Fluorescence dyes suited for indication of environmental conditions are immobilized inside this molecular multishell. The low thickness of about 10 nm of this surface films ensures, that the indicator beads respond very fast on chemical changes in the surrounding media.

Dye labelled polymer particles (Rhodamin G embedded in a polyelectrolyte layer on a polystyrene particle) were included in investigations for the labelling of segments. In contrast to indicator beads, therefore fluorescence-labelled particles are needed, in which the response of fluorescence intensity on chemical change of the segment liquid can be neglected. Polymer molecules with an inert matrix completely filled with immobilized dye are suited for this purpose (fig. 3).

In principle, label beads and indicator beads can be applied together in the same experiments. For experiment designs, it is desirable to equip each segment with the same combination of indicator beads and a varied combination of label beads (fig. 4). The equal content of indicator beads (sensor particles) ensures, that the same chemical parameters can be read-out from all fluid segments. In contrast, different combinations of labelling particles give the segments an individual signature and make the identification of a certain segment after some process steps easy. The application of a standard set of particles into segments can be realized by constant supply of bead suspensions and constant mixing, segment generation and injecting conditions. More difficult is the controlled introduction of certain label bead combinations with differences from segment to segment. In alternative, therefore a statistical approach can be chosen.

3. Results and discussion

In general, different particles can be introduced into segments. Beside the mechanical properties of them, the chemical affinity, the swelling and solvation properties of particles are influencing the behaviour inside micro fluid segments. In some cases, particles can be accumulated inside the fluid interface [16]. The application of particles with high surface affinities to the walls or to the carrier liquid must be avoided. In the experiments, carried out here, charged particles are applied which possess a high surface charge and a high affinity against the aqueous phase.

A second problem is the homogenous distribution of particles inside the solution. Small density differences between particles and liquid can cause sedimentation. The speed of sedimentation increases with increasing size of particles and increasing density differences. The effect of separation by sedimentation can be suppressed to some extend by moved fluid actuators or by special devices for stirring during actuation [15]. A better solution is the adaptation of density of particles and liquids. It could be shown, that a homogenous distribution of sensor particles inside one segment and a constant concentration of particles inside as sequence of segments can be achieved, if the density of particles is equal of the density of segment liquid. In this case, the concentration of particles and, therefore, the average number of particles per segment can be well controlled. Fig. 5 shows two examples of segments generated from density-adapted polyelectrolyte multishell particle suspensions. The image was taken from a segmented flow inside a microlithographically patterned glass channel chip. The density adaptation was achieved by mixing water with an aqueous solution of carbohydrates. A homogenous distribution of particles was achieved at high (about 1000 particles per segment) as well as at lower concentrations (about 50 particles per segment). The fluorescence imaging of segments is possible with a fluorescence microscope at low magnification. The shape of segment can clearly recognised in case of higher particle concentration (fig. 5a). The carrier liquid does not contain fluorescing particles. The number of particles adsorbed at the wall of the micro channel is low. At low particle concentration, it is difficult to detect visually the interface between process liquid and carrier liquid from the fluorescence image (fig. 5b).

Miniaturized toxicological or drug screenings, automated combinatorial synthesis or highly parallelized analytical procedures demand for the possibilities of fast automated measurement. Here, the online measurement of integral fluorescence intensity was used for the recognition and characterization of the segments. Therefore, a first experiment the micro channel chip was mounted on a microscope stage. The optical signal was collected by a lens mounted inside the camera adaptor. The light was then conducted through a glass fiber to a photodetector. This way, it was possible to measure the changing intensity of fluorescence during the passage of segments through the optical window.

Beside the measurement in micro channels of glass chips which have higher optical quality, there was the desire of characterizing also segments inside PTFE tubes. Therefore, the segment sequences inside the tubes were also conducted over the stage of a fluorescence microscope and the fluorescence light was also measured time-resolved by use of the fiber arrangement. The experiments showed that it is possible to recognized the bead-labelled segments by the fluorescence from the beads (fig. 6). High signal to noise ratios were obtained in the recognition of segments at higher concentrations of dye-labelled multishell particles (about 10 000 particles per microliter, corresponding to about 10^3 particles per segment). But, the segments can also be identified and a quantitative estimation of bead content or fluorescence activity is possible in case of low bead concentration (about 250 particles per microliter, corresponding to about 25 particles per segment).

Core-shell particles with integrated fluorescence dyes can be used as nanotransducers for the concentration measurements of substances with quenching activity. We tested the applicability of Rhodamine-containing polyelectrolyte polymer particles for the determination of pyrocatechol in alkaline aqueous solution. Fig. 7 shows quencher-sensitive particles inside a micro fluidic segment. The particles respond comparatively quickly on the application of a pyrocatechol-containing solution. At a concentration of 0.033 mmol/L we observe a decrease of fluorescence intensity within five minutes (fig 8). The fluorescence disappears much faster in case of higher concentrations The reduction of fluorescence intensity is strongly related to the applied concentration of the analyt. The quenching effect can be read-out by integral fluorescence measurements or locally by fluorescence imaging. We use a small set of particles for estimation of the quencher concentration. The quenching effect can be derived by averaging the intensity loss of several particles. The intensity values are taken from the grey levels of the fluorescence images in one wave length channel (fig. 9). So, this technique allows a certain quantification of analyt by measuring their concentration-dependent quenching activity inside nanoliter volumes.

Conclusions

The experiments showed, that polymer sensor or label beads of 3 µm, up to 20 µm diameter can be introduced into the nanoliter fluid segments with different, but well-controlled concentrations. Homogenous distributions were achieved by injecting bead dispersions in aqueous carbohydrate solutions into perfluorooctane using PTFE tube and injector material. The obtained fluid segments of about 200 nl volume contained between 10 and 1000 sensor beads depending of original concentration in the particle suspension.

The read-out of optical information can occur either by microscopic imaging or by integrating measurement of fluorescence intensity in a certain spectral range. The first method is well applicable for the characterization of segments with sensor beads inside micro channels with cover chips with plane parallel optical transparent walls. The second method is well suited for the characterization of sequence segments flowing through micro tubes. The results demonstrate the applicability of polymer-based sensor micro particles with polyelectrolyte shells for monitoring purposes in nanoliter fluid segments.

Some boundary conditions should be taken in mind for the further research on the application of sensor and label bead combinations. At first, an uncontrolled aggregation of beads must be avoided in order to get well defined bead combination inside the fluid segments. At second, libraries of labelling particles and sensor beads with the a larger spectrum of sensor functions and stepwise varied densities adapted on process solution with different substrate and salt composition must become available. At third, the optical read-out signal from the labelling particles must not interfere with the optical signal of sensor particles. This is of particular importance in relation to the fact, that chemical concentrations are normally read-out by intensity changes in fluorescence signal of sensor beads.

The spectral separation of emission light of different indicator and label particles is essential for an integral measurement of fluorescence intensity from segments. The only alternative would consist in time-dependent measurements (fluorescence decay) instead of intensity measurements. But this measurements demands for expensive equipments and seem not be applicable for a larger applications of micro fluidic compact systems at the present state of development. Much better conditions exist for the read-out of segment-related information by fluorescence imaging. Specific information can be gained, if fluorescence images of one and the same segment were taken in several optical channels (emission wave lengths) at the same time. So, the fluorescence intensity of single beads can be determined. Overlapping of fluorescence bands of different beads is no longer important for the signal quality and for the precision of intensity measurement of one type of beads.

The intensity of single beads can be brought into relation to each other if fluorescence imaging methods are applied. In addition to absolute fluorescence measurements, an internal referencing becomes possible. A disadvantage of read-out by fluorescence imaging of segments is given by the demands on optical conditions. Either the particle in the segment liquid must be focussed in a fixed optical plane for imaging or images must be taken from several focus levels in order to catch all segment particles in all local positions. In addition, special automated procedures for image processing are needed for automated screening procedures with series of segments. The fast process in development of optical systems and image processing software give the hope, that imaging methods will be available more and more easy during the next year. So, the strategy of read-out of segment information by in-situ transduction by sensor or label particles inside the micro fluid segments seems to be a very promising way.

Acknowledgement

We thank F. Möller and S. Schneider (Ilmenau) for technical assistance. The support by T. Henkel and J. Albert (supply of chip devices from the IPHT Jena) as well as valuable discussion and support by A. Groß (Ilmenau) and financial support by the German Environmental Foundation (DBU) are gratefully acknowledged.

References

[1] J. R. Burns and C. Ramshaw, Lab on a Chip 1 (2001), 10
[2] T. Nisisako, T. Torii and T. Higuchi, Lab on a Chip 2 (2002), 19
[3] I. Shestopalov, J.D. Tice, R.F. Ismagilov, Lab Chip, 4 (2004), 316
[4] J.M. Köhler, Th. Henkel, A. Grodrian, Th. Kirner, M. Roth, K. Martin and J. Metze, Chem. Engn. J. 101 (2004), 201
[5] M. Curcio and J. Roeraade, Anal. Chem. 75(2003), 1

[6] B. Zheng, J. D. Tice and R.F. Ismagilov, Anal. Chem.,76 (2004), 4977
[7] K. Martin, T. Henkel, V. Baier, A. Grodrian, T. Schön, M. Roth, J.M. Köhler and J. Metze, Lab on a Chip 3 (2003), 202
[8] A. Grodrian, J. Metze, T. Henkel, K. Martin, M. Roth and J.M. Köhler, Biosensors & Bioelectronics 19 (2004), 1421
[9] Ch. Krause et al.: Anal. Chem. 71 (1999), 5304
[10] Ch. Huber et al.: Anal. Chim. Acta 449 (2001), 81
[11] O.S. Wolfbeis, J. Mater. Chem. 15 (2005), 2675
[12] I. Estrela-Lopis, S. Leporatti, S. Moya, A. Brandt, E. Donath, H. Möhwald, Langmuir 18 (2002), 7861
[13] M. Fischlechner, I. Toellner, P. Messner, R. Grabherr, E. Donath, Angew. Chem. Internatl. Ed. 45 (2006), 784
[14] S. Hiller, A. Schnackel, E. Donath, Cytometry 64 A (2005), 119
[15] A. Brösing: Phasenverhalten von funktionalisierten Synthesepartikeln in Mikrofluidsegmenten, 3rd Workshop Chemical and Biological Micro Laboratory Techniques (Ilmenau 2006)
[16] A. Brösing, A. Funfak, J.M. Köhler, 13th Kolloquium Heiligenstadt (2006), P09

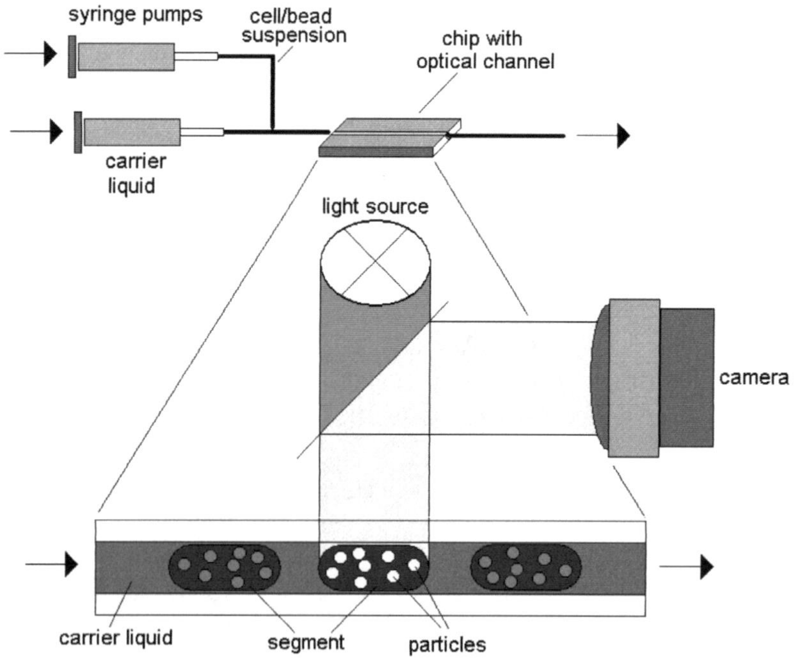

Fig. 1
Experimental arrangement for the introduction of label and indicator beads into sequences of micro fluid segments and for optical characterization of bead-labelled segments

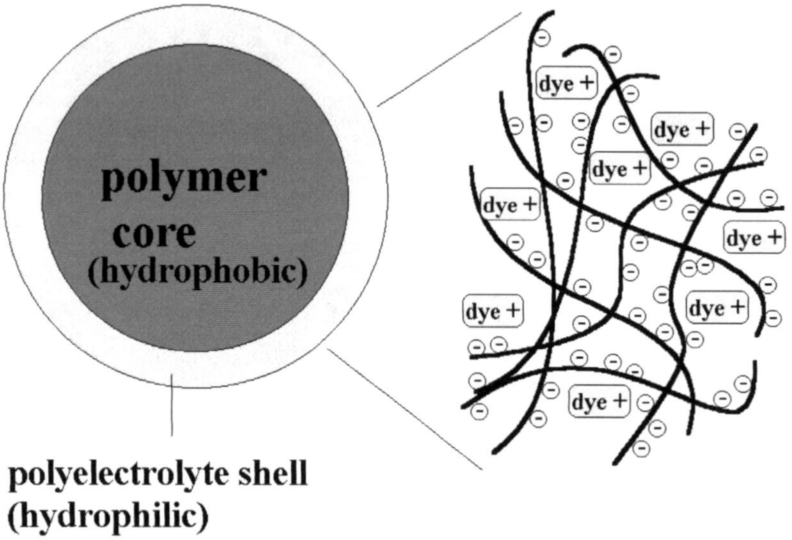

Fig. 2
a) Structure of indicator particles: hydrophobic polymer core covered by a polyelectrolyte shell incorporating the immobilzed indicator dyes, b) Multishells indicator particles generated by alternating electrostatic chemisorption of charged

macromolecules at the polymer surface

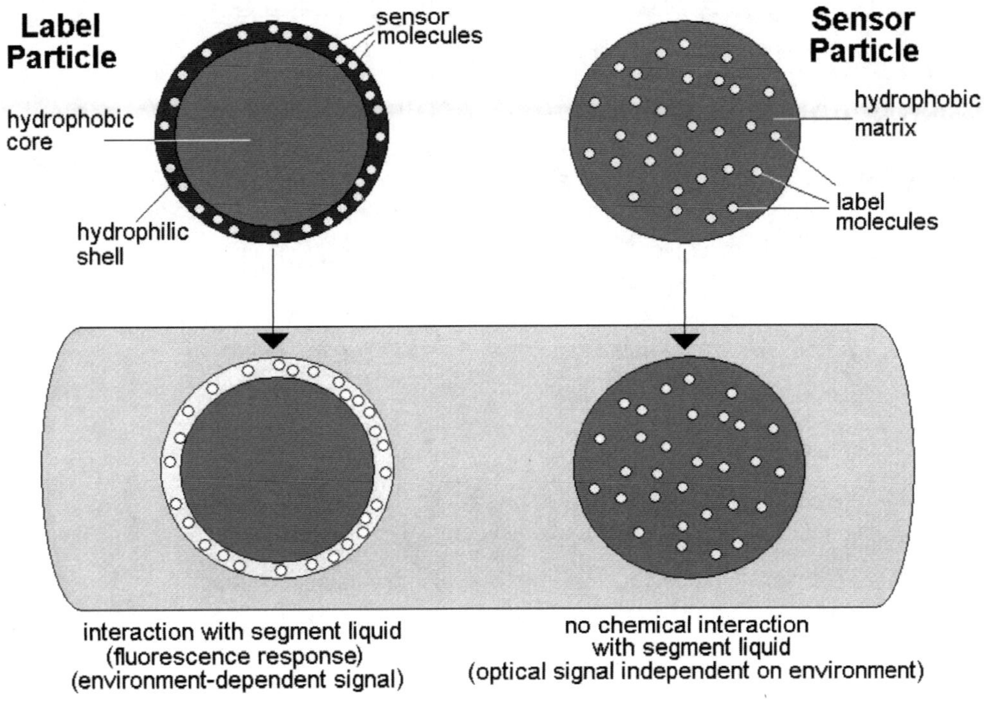

Fig. 3
Properties of label and indicator particles for segment characterization

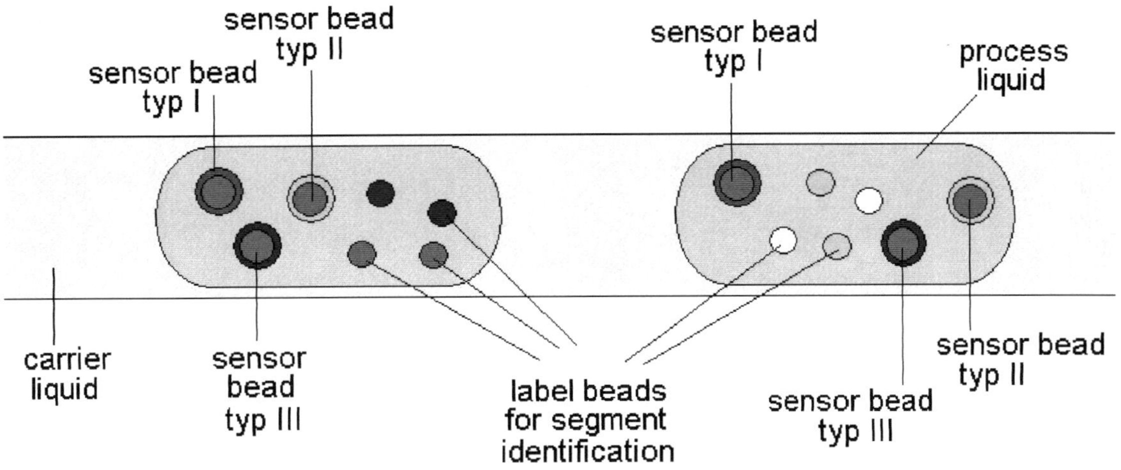

Fig. 4
Combination of label bead combinations and sensor particles in series of nanoliter fluid segments (conceptually)

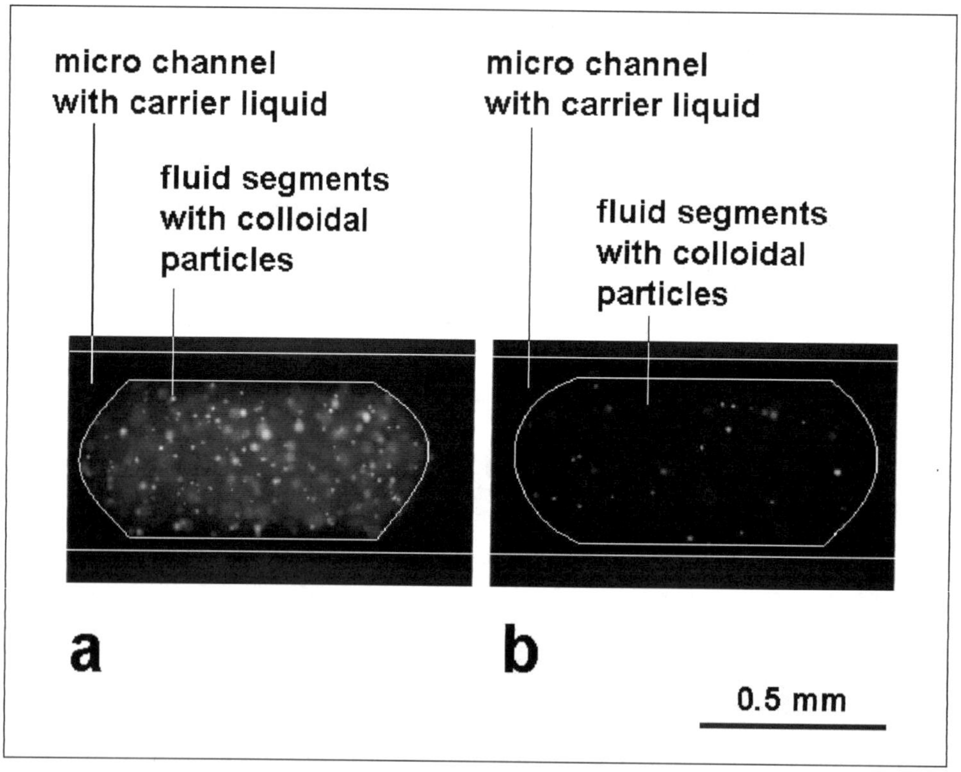

Fig. 5
Imaging of an aqueous fluid segments with sensor beads inside a lithographically prepared micro channel (emission filter: 540 nm) a) high concentration of sensor beads (about 10^3 particles per segment), b) low concentration of sensor particles, the interface between segment liquid and carrier liquid is invisible, approximative size of segment is marked by a white line.

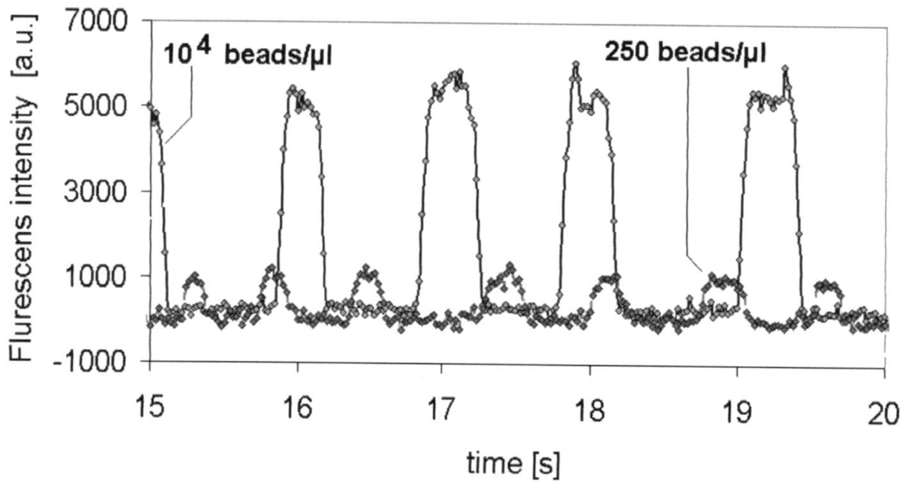

Fig. 6
Intensity-time plot for fluid segments with incorporated fluorescence-labelled micro beads flowjng through a PTFE micro tube of 0.5 mm diameter

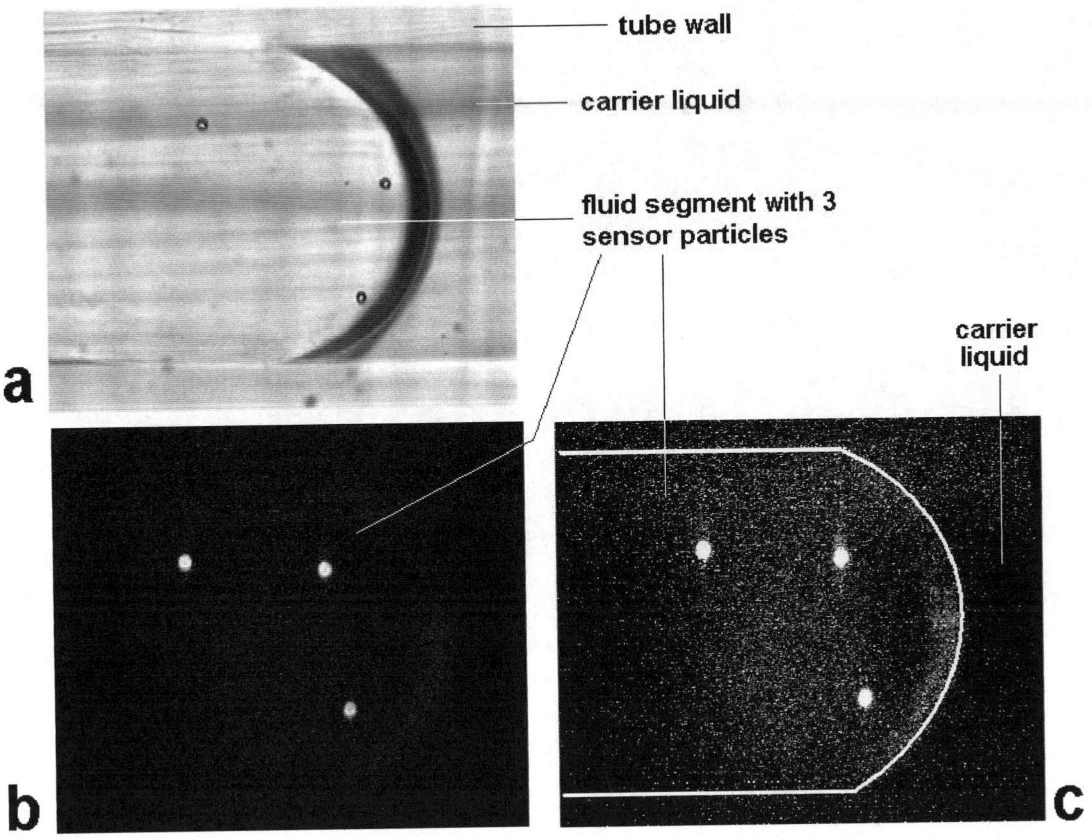

Fig. 7
Quencher-sensitive core-shell particles inside a micro fluid segment: a) transmissive light microscopic image, b) fluorescence image, c) contrast-enhanced fluorescence image (the white line marks the boundary of segment)

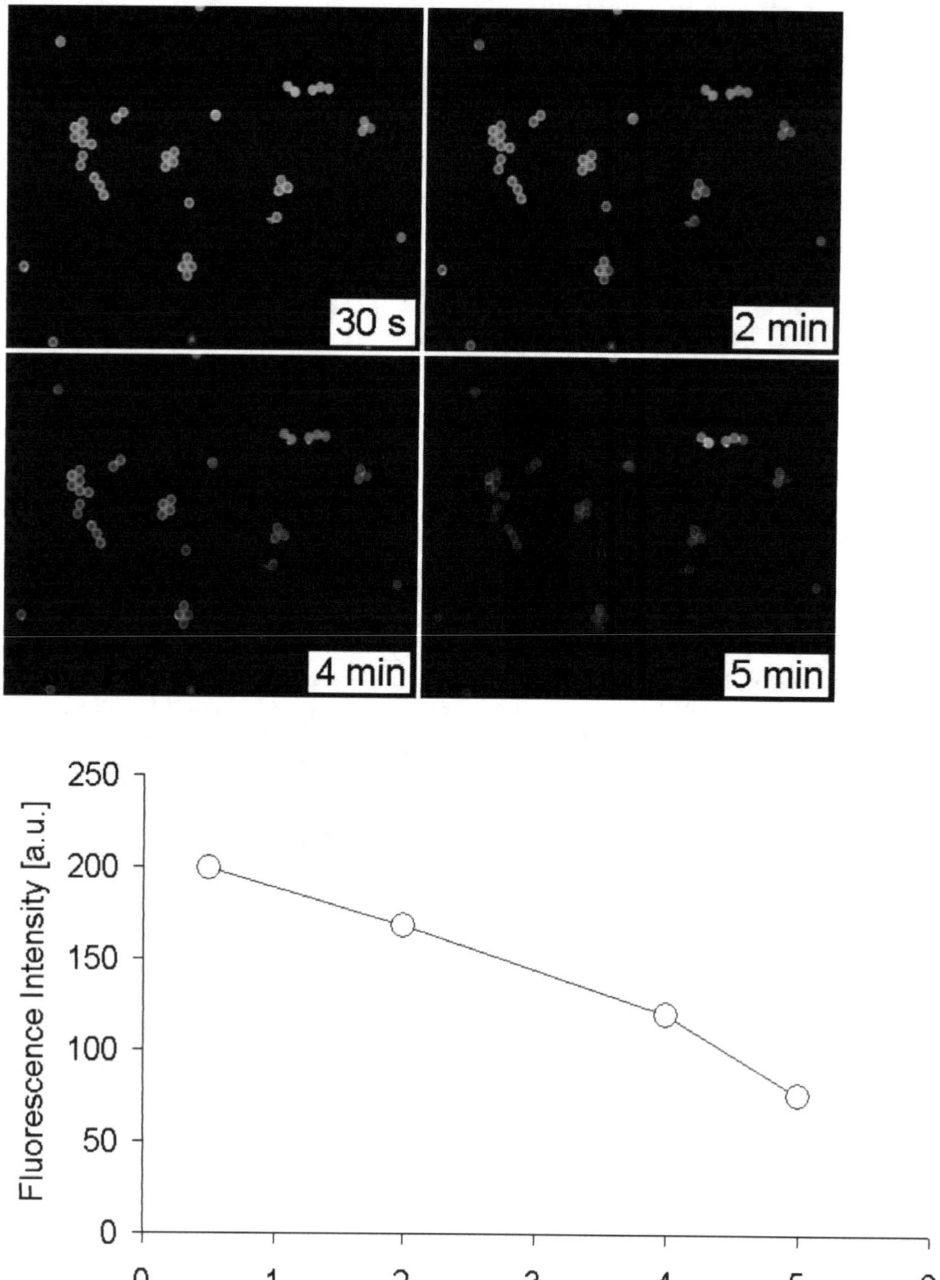

Fig. 8
Response behaviour of sensor beads: decrease of fluorescence intensity from indicator dye (rhodamine) immobilized in the shell of polyelectrolyte polymer core shell microparticles after application of a quencher solution (pyrocatechol): a) microscopic images (Particles of 20 µm diameter), b) change of fluorescence intensity over time after application of a solution containing 33 µmol/L pyrocatechol

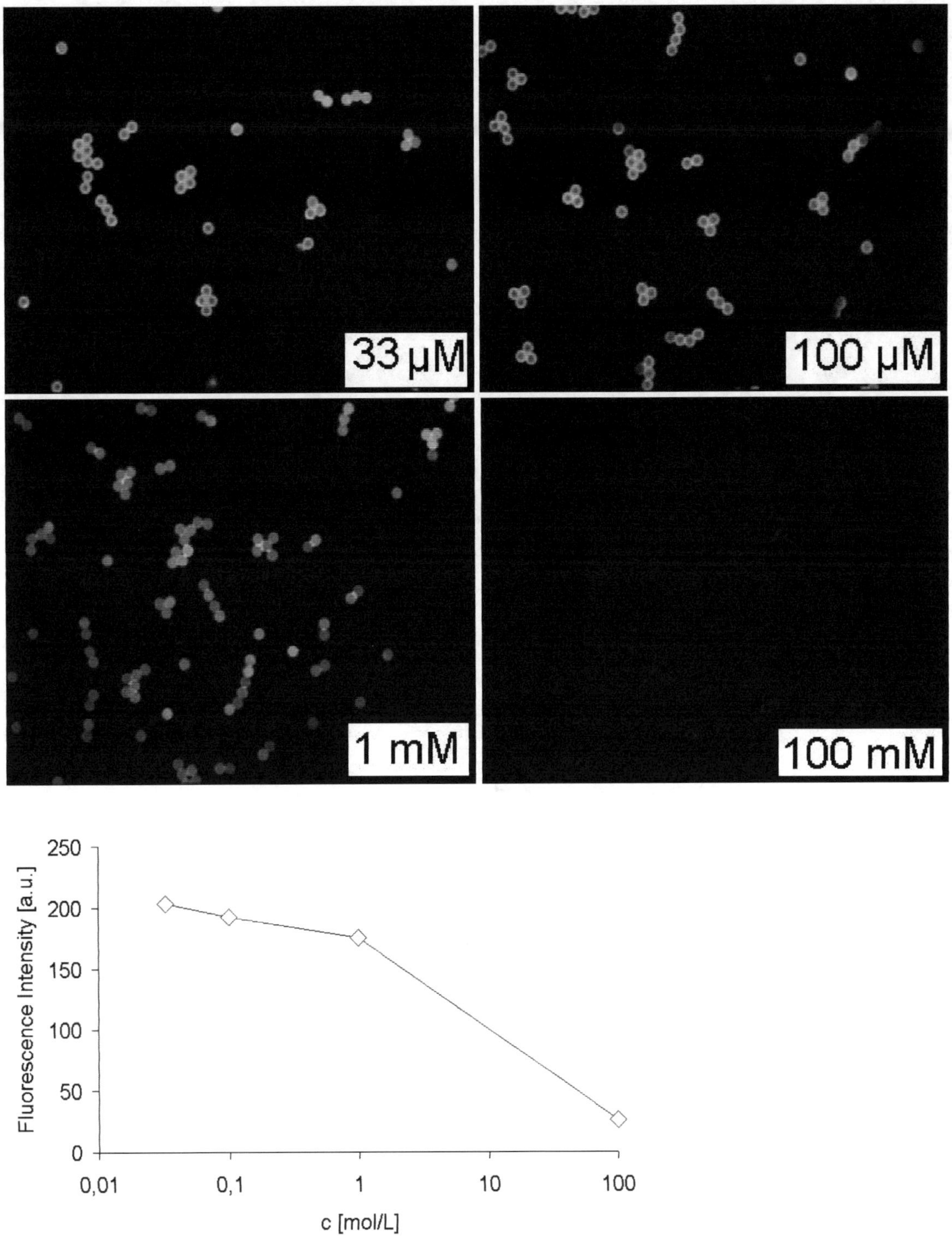

Fig. 9
Concentration-dependent fluorescence intensity for determination of analytes inside nanoliter fluid segments by measuring their quenching activity: Fluorescence images (concentrations of 0.033 mmol/L ...100 mmoll/L pyrocatechol in aqueous solution): fluorescence images, b) concentration dependence of fluorescence intensity (arbitrary units) 0.5 min after application of quencher solution

Synthesis of Nickel Crystals in Sodium Dodecyl Sulfate-Polyvinylpyrrolidone Clusters by Hydrazine Reduction

Yun FANG* Jun XU

(School of Chemical and Material Engineering, Southern Yangtze University,1800 Lihu Avenue, Wuxi, Jiangsu 214122, China)

Nickel crystals were prepared from nickel chloride by hydrazine reduction in sodium dodecyl sulfate(SDS)-polyvinylpyrrolidone(PVP) aqueous solution. The prepared nickel crystals are face centric structural and their mean size are of 16 nm, and the linear nickel crystals self-assembles into spherical particles. The size of nickel particle decreases with the increase of SDS concentration, implying SDS-PVP clusters have some size-control effect in the preparation of nickel particles.

Keywords: Polyvinylpyrrolidone; Sodium dodecyl sulfate; Soft template; Hydrazine; Nanoparticle

Introduction

Ultrafine nickel particles especially nanoparticles have novel catalytic, electrical and magnetic properties and can be used as catalyst, battery, *etc*. The finer of the catalyst's size, the higher of the catalytic activity. For example, using 30 nm nickel particles as catalyst to synthesize optical active 3-hydrobutyric methyl ester, the yield was as high as to 85% and that could speed the reaction rate 15 times higher than regular nickel catalyst did[1].

Many methods have been applied to prepare nanoparticles, in which surfactants or macromolecules are often used, such as reversed micelles[2-3], microemulsion[4-5], liquid crystalline phases[6], vesicles[7-8], aqueous macromolecule solution[9-10], block copolymers[11] and polymer gels[12]. The use of surfactant- macromolecule clusters as template to prepare inorganic crystals has become an area of explosive research interest recently[13-15]. As it is well-known water-soluble nonionic macromolecule, such as polyvinylpirrolidone (PVP) or polyethyleneoxide (PEO), interacts strongly with anionic surfactant, such as sodium dodecyl sulfate (SDS), in aqueous solution[16-18]. Above the critical association concentration (*cac*), surfactants aggregate on macromolecule chains in micellar structure, named as bound-micelle[18]. NMR proof[19-20] showed that these bound-micelles aggregate evenly on macromolecule chains.

In this paper, surfactant-macromolecule clusters were used in the preparation of nickel crystals, and hydrazine was used as reducer. XRD and SEM were used to characterize the size and morphology of the products.

Experimental Section

Materials. Sodium dodecyl sulfate (SDS, Acros Organics, 99%); polyvinylpyrrolidone (PVP K-30, ISP Co.); hydrazine, NiCl2·6H2O, Na2CO3 are all in analytical grade and obtain from Shanghai Chemistry Reagents Co. of China Medical Corp.; doubly de-ionized distilled water was used throughout.

Formation of SDS-PVP Clusters. Drop volume method[21] was used to measure the surface tension (γ) of $NiCl_2$-SDS binary systems and $NiCl_2$-SDS-PVP ternary systems to determine the formation

* Correspondence to: Yun Fang, School of Chemical and Material Engineering, Southern Yangtze University,1800 Lihu Avenue, Wuxi, Jiangsu 214122, China. E-mail: yunfang@126.com

boundary of SDS bound-micelle. The drop volume measurement was performed at (40.0 ± 0.5) °C and repeated at least 4 times until the relative error was lower than 1.0%.

Preparation and Characterization of Nickel Particles. A series of $NiCl_2$-SDS-PVP aqueous solutions with different concentration of SDS were prepared and left in thermostat at (40.0 ± 1.0) °C for 2 hr. Adjusted pH value of the solutions to about 11 by Na_2CO_3. Hydrazine was added into the solution and the molar ratio of nickel ions and hydrazine was maintained to 1:10. The reaction was carried out at 60.0 (± 1.0) °C. When the solution became gray, it means the formation of nickel particles. The formed particles remained in suspension being presumably stabilized by SDS or PVP chains. The black products were collected with centrifuge (10,000 r/min), washed with distilled water and ethanol for 3 times, respectively, and then desiccated at room temperature.

X-ray diffraction (XRD) measurement was performed with an instrument of Rigaku D/max2500 to analyze the structure of the prepared particles, and the size of particles was calculated from Scherrer formula. The nickel particles was dispersed in ethanol and characterized by Quanta-200 scanning electron microscope (SEM).

Results and Discussion

Reaction Condition. Hydrazine is a normal reducer and its reductive ability depends on the pH of the solution. In acidic medium, φ^o_{N2H4} is 0.23V and N_2H_4 is easy to be oxidized to NH_3; in basic medium, φ^o_{N2H4} is -1.16V and it is easy to be oxidized to N_2. At 25 °C, $\varphi^o_{Ni/Ni^{2+}}$ is -0.221V, so it is possible to reduce nickel ions in basic medium. The reaction equation is as following:

$$2Ni^{2+} + N_2H_4 + 4OH^- = 2Ni + N_2\uparrow + 4H_2O \qquad (1)$$

Na_2CO_3 or NaOH can be used to adjust the pH. In the experiment, Na_2CO_3 was used to adjust pH to about 11.

The Crystal Structure of Nickel Particles. Figure 1 is the XRD pattern of products prepared in the presence of SDS-PVP clusters. The 2θ at 44.5°, 51.8° and 76.4° are the characteristic absorption of nickel crystal at (111), (200) and (222), which means the prepared products are nickel crystal with face centric (fcc) structure. In the meantime, there is no characteristic absorption of NiO. It means the nickel crystals haven't been oxidized, which may be the result of SDS-PVP clusters' protection. From Scherrer equation, the mean radius of nickel crystals is 16 nm.

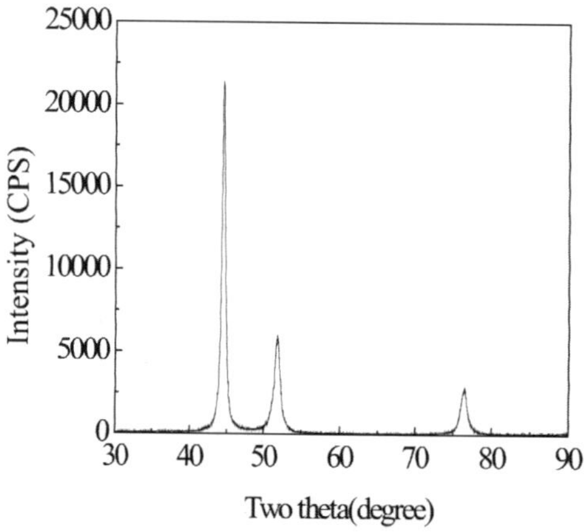

Figure 1. XRD pattern of nickel particles prepared from $NiCl_2$-SDS-PVP aqueous solution, $[NiCl_2]$=15 mmol·L^{-1}, [SDS]=10 mmol·L^{-1}, [PVP]=7 g·L^{-1}.

Figure 2 is the SEM images of nickel particles. It shows that the products are spherical nickel particles, their mean diameter ranged mainly from 0.5 to 3.0 μm, and these nickel particles are formed by many fine linear nickel crystals (ca. 20 nm in length), the whole nickel particle has some special morphology. SDS-PVP cluster has such a structure that allows SDS micelles aggregate on PVP chains, and these micelles are called as bound-micelles. SDS is a kind of anion surfactant; therefore nickel ions can associate with the head group of SDS through electrostatic interaction, so that the surface of bound-micelles can act as microreactor in the reduction process of nickel ions. Liner PVP molecule has template effect[10] and will be attracted by newborn nickel crystals, therefore the newborn nickel crystals will grow along PVP chains, which may be the reason of the formation of these linear nickel crystals. Nanoparticle has high surface energy and also nickel crystal has magnetic property, so that these fine nickel crystals attract with each other and aggregate to decrease the surface energy.

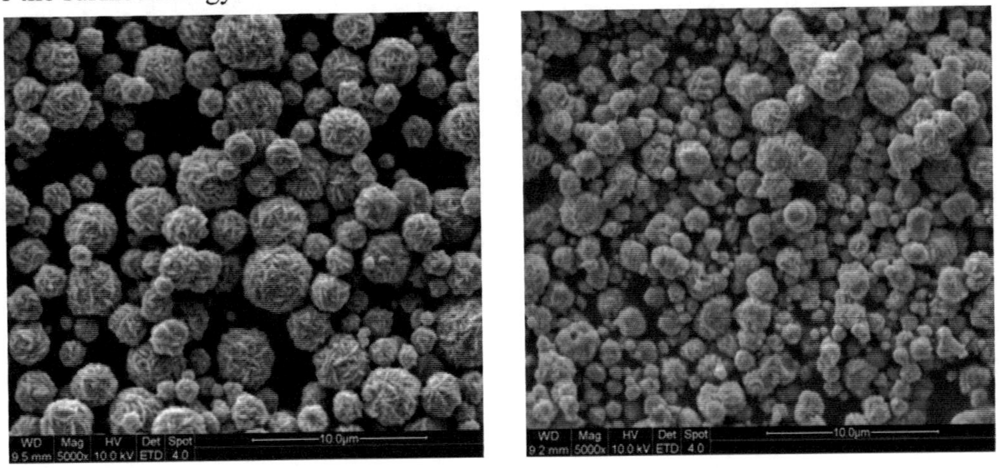

(a) (b)

Figure 2. SEM images of nickel particles, (a) [SDS]=20 mmol·L^{-1}, [$NiCl_2$]=40 mmol·L^{-1}, [PVP]=7 g·L^{-1}, (b) [SDS]=50 mmol·L^{-1}, [$NiCl_2$]=40 mmol·L^{-1}, [PVP]=7 g·L^{-1}

Effect of SDS Concentration. Figure 2(a) and figure 2(b) show that the size of nickel particles decreases with the increase of SDS concentration, implying that SDS-PVP clusters can manipulate the formation of nickel particles. With higher SDS concentration, more SDS bound-micelles form on PVP chains, which induce more nickel crystals formed at the same time, leading smaller nickel particles. As a result, the surface of the particles become large and the catalytic activity will be better. In the meantime, the nickel crystals infuse with each other because they are closer to each other, as it can be seen in figure 2(b).

Effect of Nickel Ions Concentration. Figure 3 shows that when nickel particles were prepared at lower concentration of nickel ions, big particles formed and the size distribution of particles is broader. It is because when nickel ions concentration is lower, few nickel crystals formed at the same time and lead to the increase of size distribution.

Figure 3. SEM image of nickel particles, [SDS]=10 $mmol \cdot L^{-1}$, [$NiCl_2$]=20 $mmol \cdot L^{-1}$, [PVP]=7 $g \cdot L^{-1}$

Conclusions

Nickel crystals were prepared by hydrazine reduction in SDS-PVP aqueous solution. XRD shows that the nickel crystals have fcc structure and the mean size is of 16 nm. SEM images show that linear nickel crystals self-assemble into spherical particles. The sizes of nickel particles changed with the changing of either SDS or nickel ions concentration. With the increase of SDS concentration, the size of spherical nickel particles decreases, implying the template effect of SDS-PVP clusters.

Acknowledgment. We thank the financial support provided by the National Science Foundation of China (20371021).

References

(1) Harada, T. S.; Tai, O. A. *Chem. Lett.* **1977**, *10*, 1131.
(2) Kimijima, K.; Sugimoto, T. *J. Colloid Interface Sci.* **2005**, *286*, 520.
(3) Petit, C.; Lixon, P.; Pileni, M. P. *J. Phys. Chem.* **1993**, *97*, 12974.
(4) Zhang, K.; Chew, C. H.; Xu, C. Q. *Langmuir* **1999**, *15*, 3056.
(5) Gao, L.; Wang, E.; Lianm S. Y. *Solid State Commun.* **2004**, *130*, 309.
(6) Scott, R. J. Oliver; Geoffrey, A. O. *J. Mater. Chem.* 1**998**, *8*, 1081.
(7) Kennedy, M. T.; Korgel, B. A.; Monbouquette, H. G. *Chem. Mater.* **1998**, *10*, 2116.
(8) Zheng, X. W.; Xie, Y.; Zhu, L. Y.; etal. *Ultrason. Sonochem.* **2002**, *9*, 311.
(9) Marignier, J. L.; Belloni, J.; Delcourt, M. O. *Nature* **1985**, *317*, 344.
(10) Zhou, K.; Zhang, X. H.; Wu, S. K. *Acta Chimica Sinica* **2004**, *62*, 1771.
(11) Chiu, J.J.; Kim, B.J.; Kramer, E.J.; Pine, D.J. *J. Am. Chem. Soc.* **2005**, *127*, 5036.
(12) Breulmann, M.; Davis, S.A.; Mann, S.; Hentze, H. P. *Advanced Materials* **2000**, *12*, 502.
(13) Nizri, G.; Magdassi, S. *J. Colloid Interface Sci.*, **2005**, *291*, 169.
(14) Cheng, T.; Zheng, S.; Helmeuth, M. *Langmuir* **2003**, *19*, 9039.
(15) Maria, O.; Epameinondas, L.; Tasoula, K. L. *Langmuir* **2004**, *20*, 5605.
(16) Goddard, E. D. *Colloid Surface A* **1986**, *19*, 255.
(17) Zana, R.; Linaos, P.; Lange, J. *J. Phys. Chem.* **1985**, *89*, 41.

(18) Fang, Y.; Cai, K.; Zong, L. Y.; Liu, X. F.; He, M. S. *Chem. J. Chinese U.* **2004**, *25*, 888.
(19) Fang, Y.; Liu, X. F.; Xia, Y. M.; Zong, L. Y. *Chem. J. Chinese U.* **2006**, *27*, 731.
(20) Cabane, B.; Duplessix, R. *J. Phys.* **1987**, *48*, 651.
(21) Zhao, G. X.; Zhu, B. Y. *Principles of Surfactant Action*; China Light Industry Press: Beijing, **2003**.

Mechanical Oscillation of Pnipam Gel Particles Induced by Belousov-Zhabotinski Reaction

Zhengdong Cheng, Texas A&M University, Chemical Engineering Department, College Station, TX 77843, Srinivasa Pullela, Artie McFerrin Department of Chemical Engineering, Texas A&M University, College Station, TX 77843-3122, Jingyi Shen, Artie McFerrin Department of Chemical Engineering, Texas A&M University, 409 Jack E Brown Building, Texas A&M University, College Station, TX 77843-3122, and Manuel Marquez, Arizona State University; National Institute of Standards and Technology; Los Alamos National Laboratory.

Uniform PNIPAM gel particles with covalently bound tris(bipyridyl)ruthenium(II) were synthesized via coordination chemistry between a ruthenium complex and the monodispersed poly(N-isopropylacrylamide) (PNIPAM) microgel particles bearing bipyridine ligand. UV-Vis spectroscopy and confocal microscopy were employed to confirm the loading of the tris(bipyridyl)ruthenium(II) complexes and the uniformity of the particle size. Periodic mechanical oscillations induced by BZ reactions were observed in these PNIPAM gel particles. A systematic study was conducted to investigate the effect of BZ substrate concentrations on the oscillation behaviors.

An Overview of Microtechnology in Process Analytics

Melvin Koch, Center of Process Analytical Chemistry, University of Washington, 160 Chemistry Library Building, Box 351700, Seattle, WA 98195-1700

Process Analytics has been migrating toward greater use of micro technology for some time to improve the performance of on-line instrumentation. This presentation will describe how the advantages of the approach impact the ability to use process instrumentation in on-line applications. It will then focus on how the instrumentation changes have enabled process analytics to be used for advantage in the characterization of microscale processes. The description will span the range of applications from lab scale to production scale and how the technology aids the transition. Finally, the presentation will describe the efforts being made in the field to educate others in the use of the technology.

Optical and electrochemical residence time-characterization of LTCC flow-through microreactors

G. Alexander Groß[1,2)], Lahbib Abahmane[1,2)] , Torsten Thelemann[1)], Michael Fischer[1)], Michael Hinz[1)], Dusan Boskovic[3)], P.M. Günther[1,2] , J. Michael Köhler[1,2)]
[1)] Techn. Univ. Ilmenau, Institut for Micro and Nanotechnologies, Germany
[2)] Techn. Univ. Ilmenau, Inst. for Physics, Germany
[3)] Fraunhofer-Inst. for Chemical Technology, Pfinzthal, Germany

correspondence:
Prof. Dr. Michael Köhler, Technische Universität Ilmenau, Institut for Micro and Nanotechnologies, Institut for Physics, PF 10 05 65, D- 98684 Ilmenau, Germany, Tel. 49- (0) - 3677 - 69 - 3655, Fax - 69 - 3179 , michael.koehler@tu-ilmenau.de

Abstract

A modular microreactor system for reactions at elevated temperature and pressure was developed using the technology of low-temperature cofired ceramics (LTCC). The technology allows the integration of heating elements as well as electrochemical transducers. A set of microelectrode pairs was applied for the determination of the residence time behaviour of different parts of the microreactor system. Conductivity measurements of residence timewere compared with optical measurements using external optical sensors at the inlet and outlet. The results reflect a good agreement between both methods at lower flow rates, but strong differences in the monitoring of global and local residence times at high flow rates.

Keywords

Microreactor, Residence time distribution (RTD), Conductometry, Photometry, Ceramics, LTCC

Introduction

The choice of materials for microreactors depends on the special demands of the applications as well as on the preparation technology of the devices [1]. Whereas polymer material is widely applied for biological and a lot of analytical applications, more chemical and temperature resistive materials like steel, glass and silicon are applied for chemical synthesis at elevated temperature and for processes with corrosive chemicals. Beside classical micro technical materials, ceramics are well suited for chemical devices due to their extraordinary high robustness against nearly the whole spectrum of chemicals (with exception of strongly alkaline liquids and hydrofluoric acid) and their high temperature resistance [2].

The characterization of residence time behaviour is of high importance for all reactors [3] and in particular for ceramic microreactors, because this type of reactors can not be monitored by visual inspection or microscopic imaging due to the non-transparency. The determination of residence times and residence time distribution is more difficult in micro reactors in comparison with classical reactors in cause of their low volumes [4-7]. A characterization of residence time distribution of a certain microreactor or a complete microfluidic arrangement demands for the registration of tracer profiles before and after passage of the device which should be characterized [8]. The interpretation of changes in tracer profiles during their move through the reactor can be supported well by theoretical description of convective and diffusive transport of tracer inside the fluidic system. Optical techniques and integrated sensors are used for process analytics in microreators [9]. Here, the application of photometric and of conductivity measurements for the characterization of residence times and residence time distribution of ceramic reactors prepared in LTCC technology is discussed.

Experimental

The reactors are constructed in a multilayer scheme using the technology of Low Temperature Cofired Ceramics (LTCC). The raw material consist of plates of uncured ceramic, so-called green-tape. It contains the inorganic components of the ceramics which should be formed during the firing process and an organic binder. In addition to the dielectric material components, some additives can be included in this tapes for integration into devices. So, printed pattern of metal containing films (for fabrication of micro heaters, sensor elements, thermoresistive temperature sensors, e.g.) or for electrodes (for conductometric measurements or electrochemical processes) can be applied. The raw tapes can be structures mechanically in a first step (fig. 1). So the needed elements of a certain channel structure, alignment holes, vertical fluid connections, and so on, are generated in a set of green tapes. The printing of metal films occurs after mechanical patterning of the raw substrates, normally. After finishing of structuring and coating of the green tapes, this tapes are assembled into a stack in the right order for forming the fluid channels and the electrical interconnections. Therefore, all layers must be aligned in x- and y- direction and in rotation angle. Alignment holes and alignment sticks are used for this procedure. Then the laminated stack of substrates is prepared for firing. During the firing, the grains of inorganic materials are connected and the rigid ceramic matrix is formed. The organic binder material is loosed by oxidation. The resulting volume reduction is accompanied by certain shrinkage of devices. In the finishing phase, the LTCC substrate stack can further equipped with holes for fluid ports or other structures. Therefore, laser drilling is preferably applied. Finally, the fired substrate stack can be cut into single devices.

The multilayer principle of LTCC technology allows the fabrication of complex channel topologies as well as the inclusion of complex patterns inside channel structures. So, it is possible to integrated structure elements for twisting of laminar flows. This can be used for improving the mixing of liquids by interdiffusion. The integration of a series of micro beams for twisting of fluid lamellas in a meandering micro channel results in a flow through mixer structure (Fig. 2). The devices show a much better mixing performance than the channel without the micro beam structures.

The LTCC devices are marked by two important advantages: Firstly, functional elements of different metal materials can easily be integrated without expensive vacuum deposition and microlithographic steps. Secondly, the technology allows the fabrication of complex substrate geometries by application of up to twelve or more layered green tape substrates. In addition, the LTCC devices possess after firing a high chemical and thermal stability comparable with the robustness of silicon and glass reactors. Disadvantages are the comparatively high roughness of channel walls and the non-transparency of the material. The last is the reason, why an optical monitoring inside channels is not possible.

Residence time measurements based on flow through photometric measurements can be performed only by optical measurements outside the inlet of the LTCC reactor and after the outlet. For optical RTD measurements, an experimental arrangement consisting on two syringe pumps for process liquid (model: water) and a label liquid (dye solution), an injector valve and two micro flow-through photometers are used (fig. 3). This arrangement allows the characterization of distribution of labelling dye before and after the passage of liquid through the LTCC reactor. The influence of reactor on the distribution of label can be characterized by comparing the time-dependent transmission at the first and the second photometer position. The derived RTD function includes the tube pieces between first photometer and fluid inlet, the inlet port, the micro channel inside device, the outlet port and the tube piece between outlet and second photometer position.

A similar arrangement was applied for the measurement of fluid resistance. Therefore the optical flow-through devices (fig. 3) were substituted by pressure sensors at the same position. This way, the flow-dependent pressure difference could be determined.

For getting information about the internal structures of LTCC device, a non-optical local measurement principle is necessary. For its realization, pairs of micro electrodes were integrated inside the LTCC devices in order to make internal conductivity measurements. So, the time-dependent change of the concentration of a conductivity influencing label could be measured from position to position inside the microreactor under exclusion from influences of connecting tubes or the fluid ports.

Results and discussion

The LTCC devices were characterized for transport behaviour of fluids in relation to the pressure drop and the residence time. At first, the fluid resistance was determined by measuring the pressure before and after the devices at comparatively high flow rates (1.. 20 mL/min). The pressure drop increases with increasing viscosity of liquid like expected. In water, the pressure drop remained below 0.5 bar up to flow rates of 20 mL/min in case of simple meander channels (fig. 4). The fluid resistance increases considerably, if devices with integrated micro structures for twist mixing are investigated. More than 2 bar pressure drop were found for streaming water at flow rates above 17 mL/min. But, for moderate flow rates (below 6 mL/min) the pressure drops are below 1 bar both for simple channels and for static mixers also in case of highly concentrated mixtures of glycerol with water (40 Vol%). The pressure drop measurement confirms the applicability of the LTCC devices for processing of liquids in a moderate range of viscosities.

The applicability of microphotometric measurements was known from earlier investigations. So, it was of particular interest, if the conductivity measurements using the pairs of integrated microelectrodes would also be suited for the estimation of residence time behaviour. For testing, we applied a diluted sulphuric acid. Fig. 5 shows the measured electrical conductivities (dots) over time for five positions inside a LTCC micro device. The measurements reflects clearly the passage of a label pulse through the device. The height and shape of signal is changed between the sensor positions like expected. So, the measurement prove the principle applicability of integrated electrical transducers for the determination of local RTD functions inside LTCC microreactors.

The conductivity measurements were compared by calculations of RTD functions using an uniform model approximation (Fraunhofer Institute for Chemical Technology, Pfinzthal, Germany). The measurement curve of the first position (first electrode pair, No. 1) was used as start set for calculation. The following distribution functions were calculated for the following positions of sensors (No. 2..5). The calculations (lines in fig. 5) were compared with the measured values of conductivity for moderate flow rates. The measured time dependencies of the single positions correlate in good manner with the calculations. The decrease of concentration maximum as well as the enlargement of half width of the transient peak are well described by the used model. The results show on the one hand, that electrical conductivity measurements are in good agreement with expected RTD response of microreactor for lower flow rates and on the other hand, that the local RTD behaviour is well described by the applied model. So, the model reflects the transport conditions for low and medium flow rates with high accuracy in relation to the ratios of medium residence times. It supplies sufficient values for the height of tracer signals at the different positions inside the channel and provides a good description of the evolution of tracer profile over the time. The comparison of simulations with the electrically measured tracer profiles confirms the regularity of signal formation at medium and lower flow rates and prove the applicability of conductivity sensing as device-internal method for the determination of local RTD functions.

More influence of the measurement principle is expected for higher flow rate. In this case, the character of laminar flow will stronger effect the formation of signals during the measurements and will be stronger dependent on the relation between lateral concentration distribution and the propagation of the label transitions. In photometric measurements, the signal is formed following Lambert-Beers law (E = eps*c*d). The measured intensity of transmitted light is caused by the additive effect of all light tracks from the source to the detector. The intensity of transmitted light is modulated by changes in the spatial distribution of absorption. So, the detected signal is dependent on the local change of absorption with respect to the shape of measurement light ray. In case of a strong laminar flow, only a small central region shows an increase in absorbance during the first part of the traces pulse (fig. 6). With increasing time, the liquid filament of high tracer concentration in the central part of tube or channel is enlarged and occupies an increasing part of the cross section of measuring light (perpendicular directions to the light way). The convective transport of tracer is superposed by a diffusive transport in axial direction as well as by a diffusive transport in lateral direction. The effect of signal shift by diffusive effects can only be neglected in case of very high flow rates and low diffusion coefficients.

Otherwise the diffusion contributes significantly to the signal formation. Beside the evolution of tracer signal in perpendicular direction to the light way, the evolution in parallel direction to the light way must be reconsidered. This effect can described more easy do the additive effect of concentration changes on absorbance ("extinction") following Lambert-Beers law (fig. 7). It is convenient for the use of optical measurements that this law reflects a linear relation (direct proportionality) between the absorbance, the concentration and the length of light way through the absorbing liquid.

The mechanism of signal formation in case of conductivity measurements is completely different from the mechanism of optical signal formation (fig. 8). In this case, the tracer is detected by an increase in conductivity of transported liquid. In contrast to optical measurement, there does not exist a direct proportionality between tracer concentration and electrical signal. This is due to the fact, that not the electrical conductivity but the electrical resistance of liquid layers show additive behaviour for the signal formation. In case of high flow rates, the liquid filament with high tracer concentration formed at the beginning of tracer passage in the central region of channel does not influence the total of conductivity significantly. An increase of conductivity can only be measured if the resistance is lowered over the whole cross section. In result, strong changes of conductivity over several orders of magnitude must be expected during the passage of the tracer (fig. 9). Therefore, the diffusion coefficients of marker components play a crucial role in the formation of conductometric transient signal. High diffusion means a fast immigration of charge carriers from the central liquid filament into the environmental liquid and, therefore, a fast increase in conductivity in the more peripheral parts of the streaming liquid column. The protons of sulphuric acid determine the conductivity response in the case of use of this acid as tracer due to their particular high diffusivity.

The experiments show that the differences in the signal formation mechanism can not longer be neglected at higher flow rates. The height as well as the shape of transient tracer profiles run out from the expected behaviour in case of fast fluid motion. So, the time order between the last optical detector and the electrical sensors seems to be inverted. In addition, the change in the shape and size of tracer peaks contradict the assumption of a simple tracer profile.

Conlusions

Ceramic microreactors prepared by LTCC technology can be used for chemical flow-through processes at medium and high flow rates. The residence time behaviour of this devices can be characterized by optical transducers arranged before the inlet port and after the outlet port of the reactor or by electrical sensors. Microelectrodes for conductive RTD measurements can be integrated into the micro channels by the LTCC procedure easily. It is possible to integrate several pairs of microelectrodes in order to get tracer profiles for different positions inside the microreactor.

For low and medium flow rates, the differences between the tracer profiles at optical and electrical sensors can be understood well by the differences in their positions. The tracer signal is enlarged due to fluidic dispersion during the passage from the inlet to the outlet of reactor. The shift in signal height and shape from position to position inside the reactor is well reflected by the integrated electrical sensors.

The changes are well related to the expected behaviour. The experimentally determined electrical data correspond well to the tracer profiles obtained by numerical simulations.

At higher flow rates, a more complex response in the sensor signals was observed. The shift in height and shape of tracer peaks depend on the flow rate and on the measurement principle. The differences between optical and electrical transduction increases with increasing flow rates. This effect can be interpreted by the differences in the mechanisms in signal formation in relation to the propagation of a tracer related pulse by the laminar flow. The different behaviour is mainly caused by the additive behaviour of concentration-related absorbance on the one hand and the inverse proportionality between concentration-related electrical conductivities and resistances on the other hand. In consequence, it is to conclude, that flow-through photometric measurements respond quickly to the first phase of the tracer passage whereas the respond of conductivity measurements is related to a farther developed phase of tracer passage. In general, the experimental determination of residence time behaviour must carried out with respect to specific flow conditions and their interference of the measurement principle.

The investigations showed the applicability of simulations based on convective and diffusive transport of tracer during the RTD measurements at low and medium flow rates. For the further development, there is a challenge for the inclusion of the signal formation mechanisms into the models and simulations for the description of residence time behaviour and the transduction of tracer-related signals at high flow rates.

Acknowledgement

The financial support of the state of Thuringia (project Kerafema) is gratefully acknowledged.

References

[1] W. Ehrfeld, V. Hessel, H. Löwe: Microreactors (Weinheim 2000), 15
[2] V. Mengeaud, O. Bagel, R. Ferringo et al.: Lab on a Chip 2 (2002), 39
[3] E. Müller-Erwein: Chemische Reaktionstechnik (Leipzig 1998), 94
[4] V. Balakotaiah, S. Chakraborty: Chem Eng. Science 58 (2003), 4769
[5] A. Gunther, S.A. Khan, M. Thalmann et al.: Lab on a Chip 4 (2004), 278
[6] J. Yoshida, S. Suga, A. Nagaki: J. Synth. Org. Chem. Jpn. 63 (2005), 511
[7] N. Kockmann: Process Engineering Methods and Microsystem Technology, in: N. Kockmann (ed.): Micro Process Engineering (Weinheim 2006), 1
[8] M. Günther, S. Schneider, J. Wagner, R. Gorges, Th. Henkel, M. Kielpinski, J. Albert, R. Bierbaum, J.M. Köhler: Chem. Engn. J. 101 (2004), 373-378
[9] S. Löbbecke, Integration of Sensors and Process Analytical Techniques, in: N. Kockmann (ed.): Micro Process Engineering (Weinheim 2006), 249

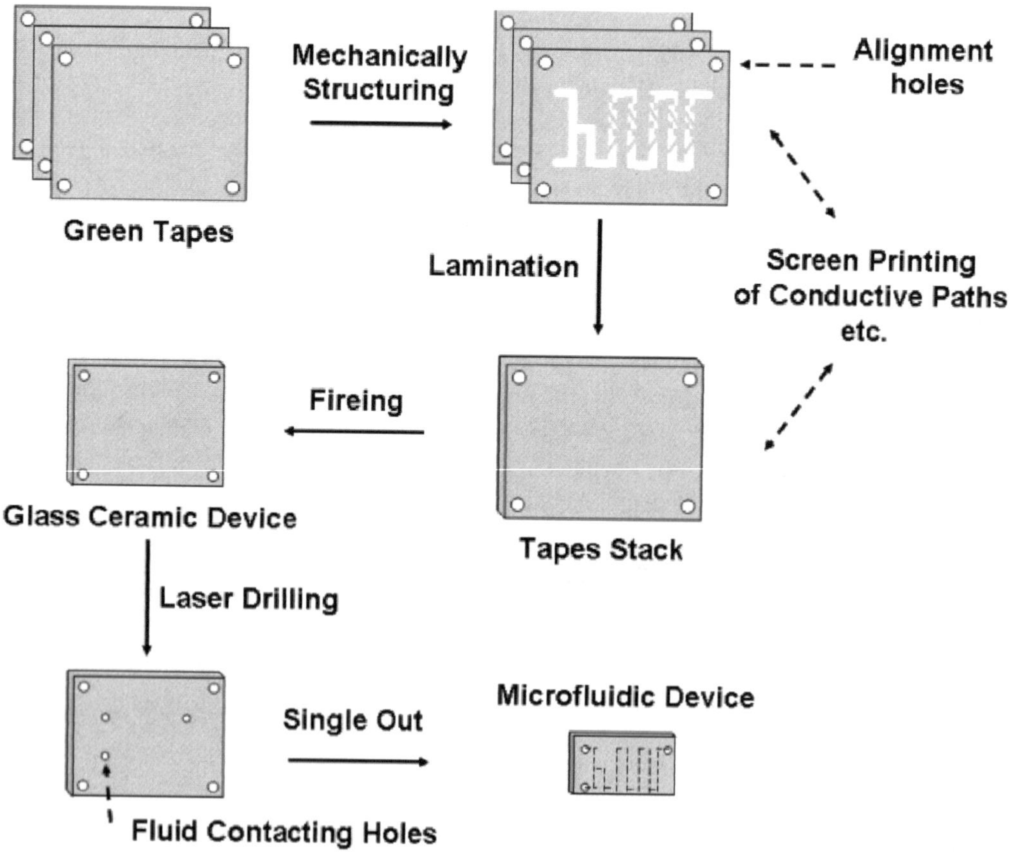

Fig. 1
Preparation of ceramic microreactors by LTCC technology

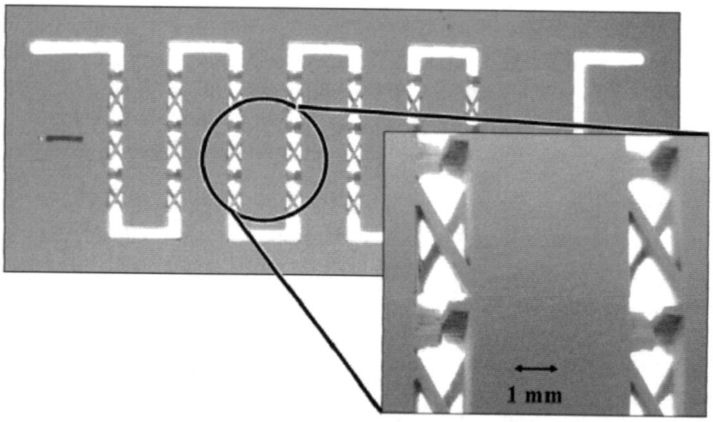

Fig. 2
Static micro mixer prepared by LTCC technology: meandering channel with

integrated micro beams for mixing by twisting fluid lamellas

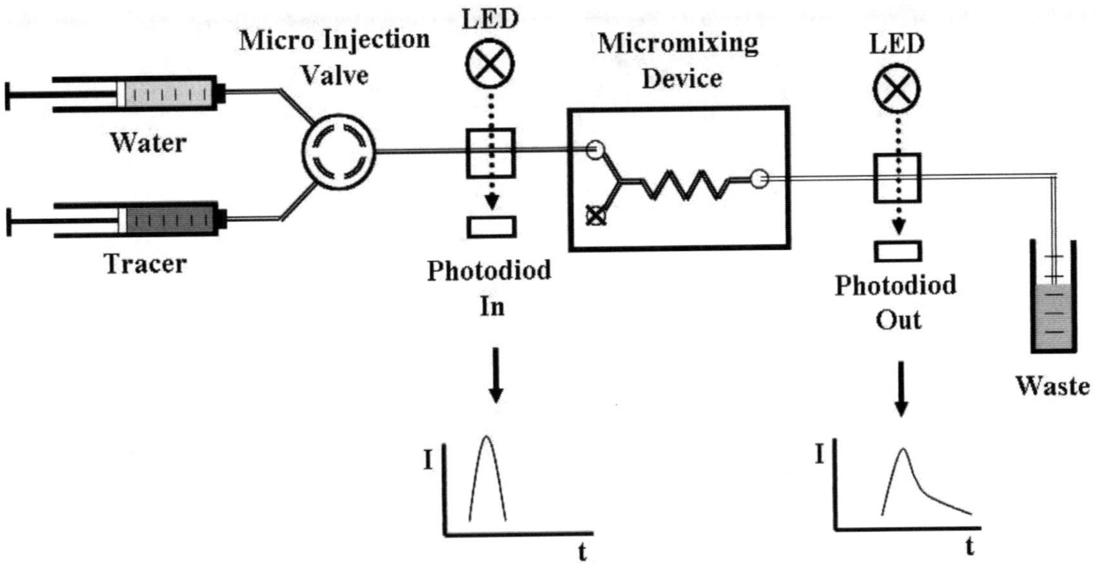

Fig. 3
Experimental arrangement for the investigation of residence time behaviour of LTCC devices by optical measurements

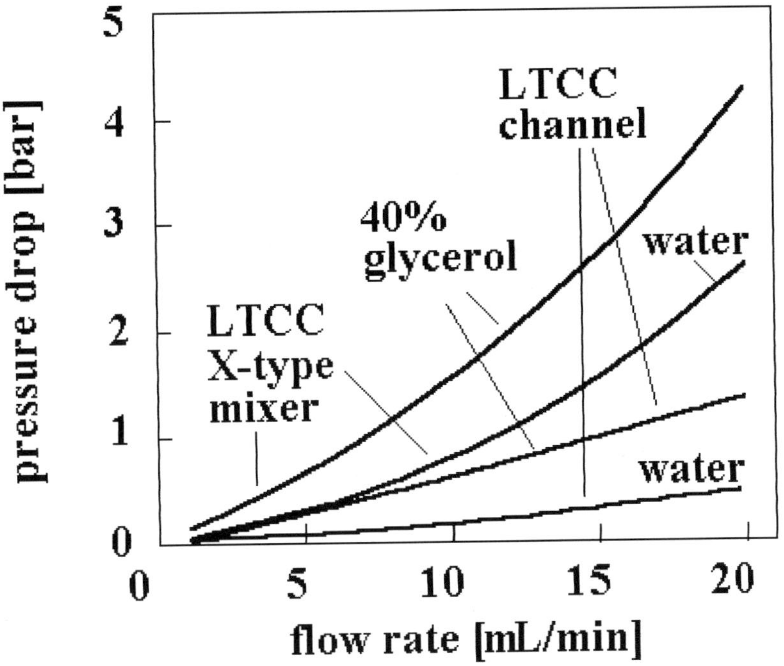

Fig. 4
Experimental determination of fluid resistance of LTCC devices for fluids with different viscosity: pressure drop in case of a simple meander channel and a X-type static micro mixer with twisting beams in dependence on flow rate; application of water and aqueous solution of glycerol

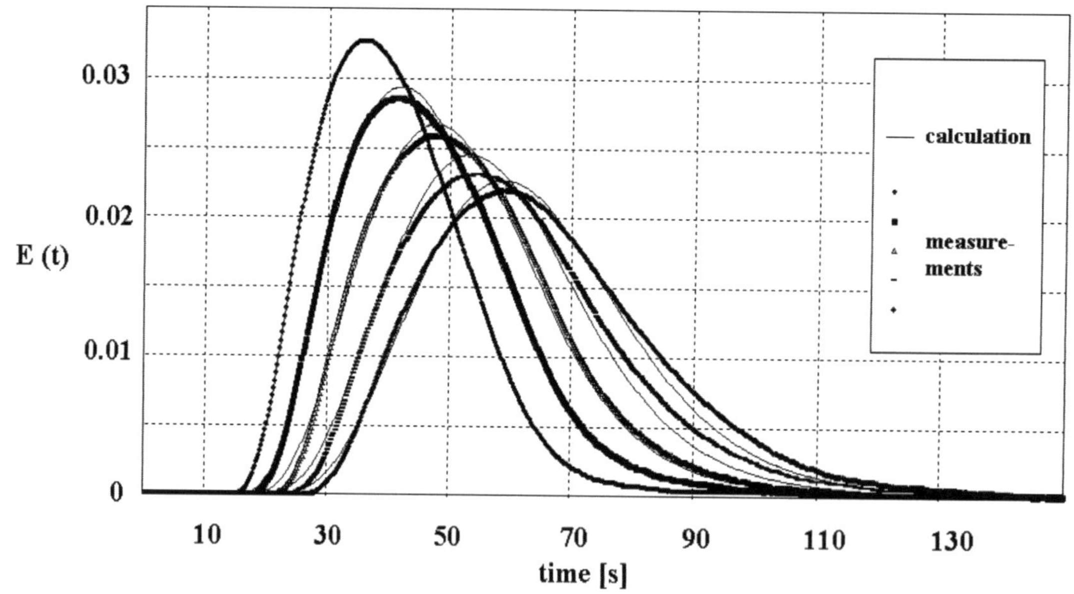

Fig. 5
Measurement of residence time behaviour of a modular LTCC microreactor system: electrical and signals at different sensor positions inside the micro channel, comparison of experimentally obtained profiles (symbols) with a RTD model (lines) for lower flow rate

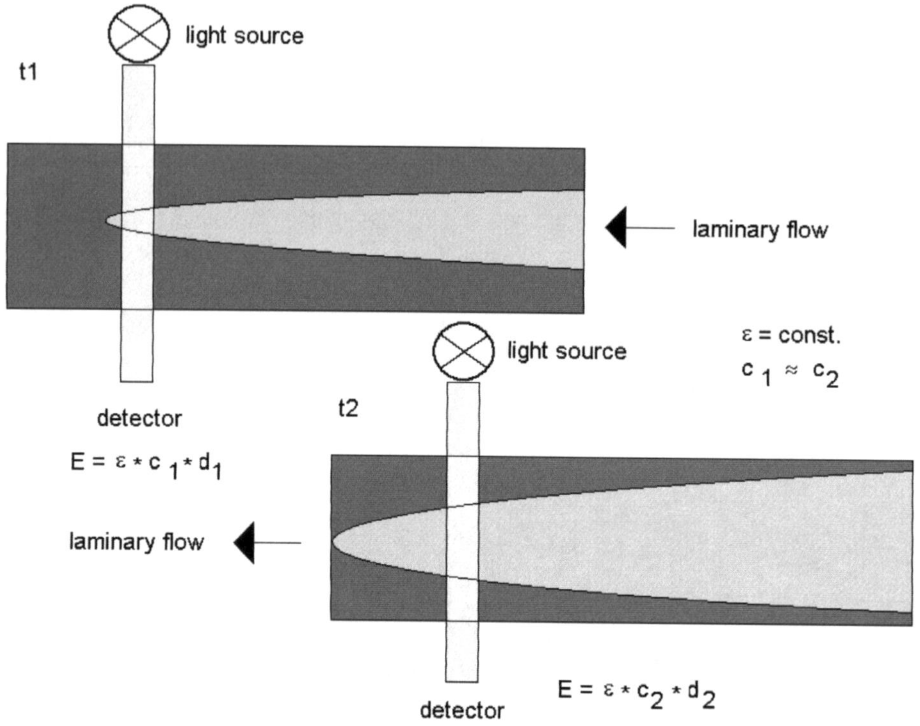

Fig. 6
Signal formation in case of microphotometric flow-through measurements (schematically)

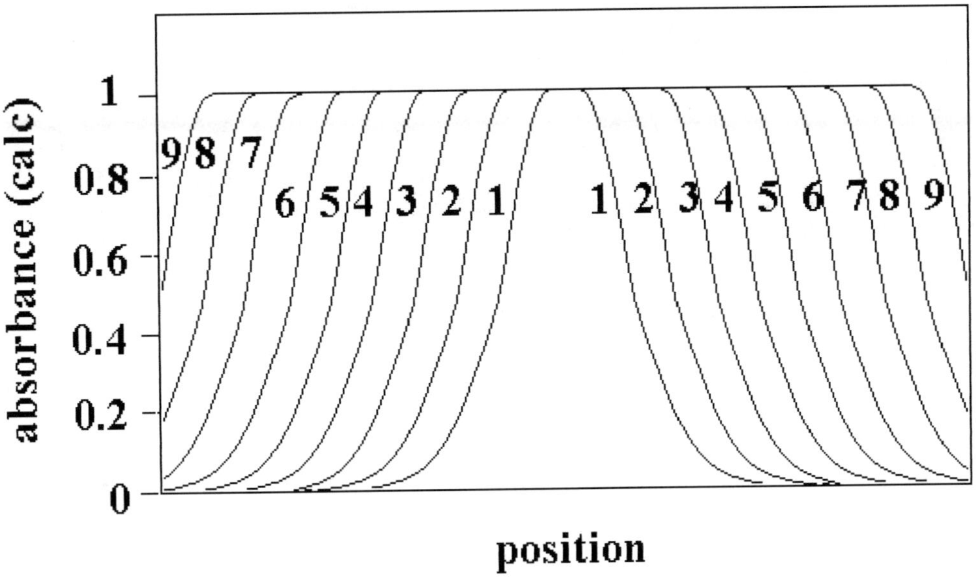

Fig. 7
Evolution of an absorbance peak at a certain sensor position in case of an optical measurement following Lambert-Beers Law in case of a laminar flow

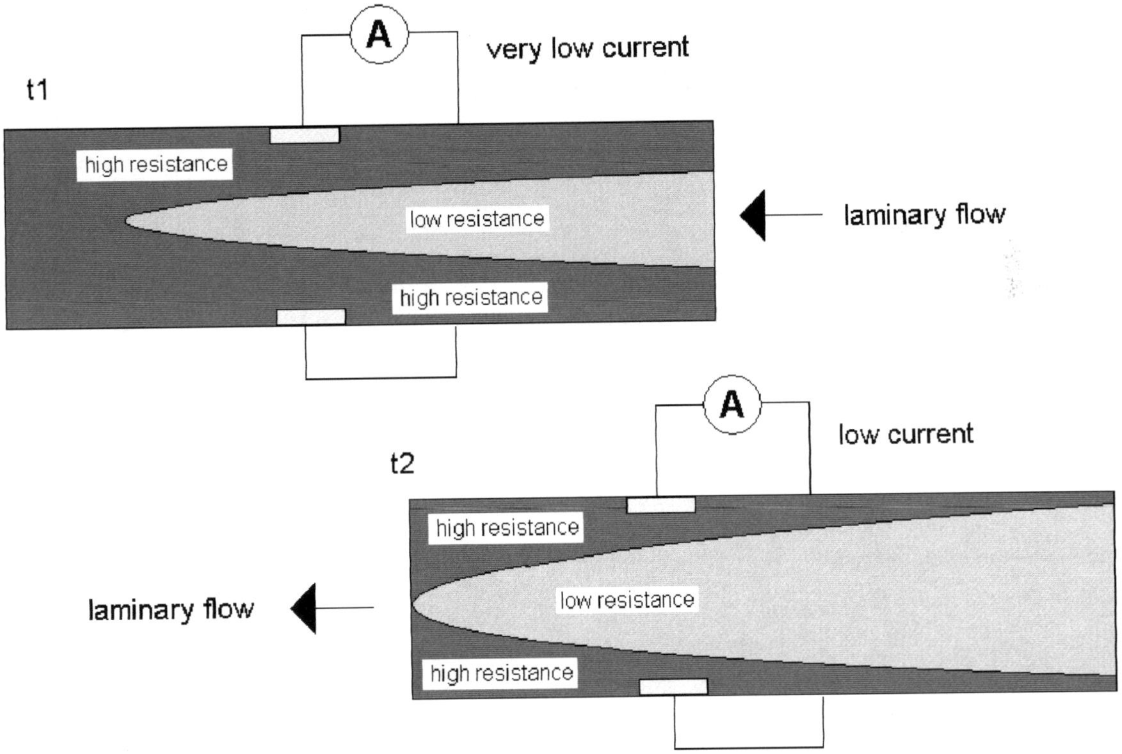

Fig. 8
Signal formation in case of a conductivity measurement with electrodes in opposite positions (top and down walls in the micro channel, schematically)

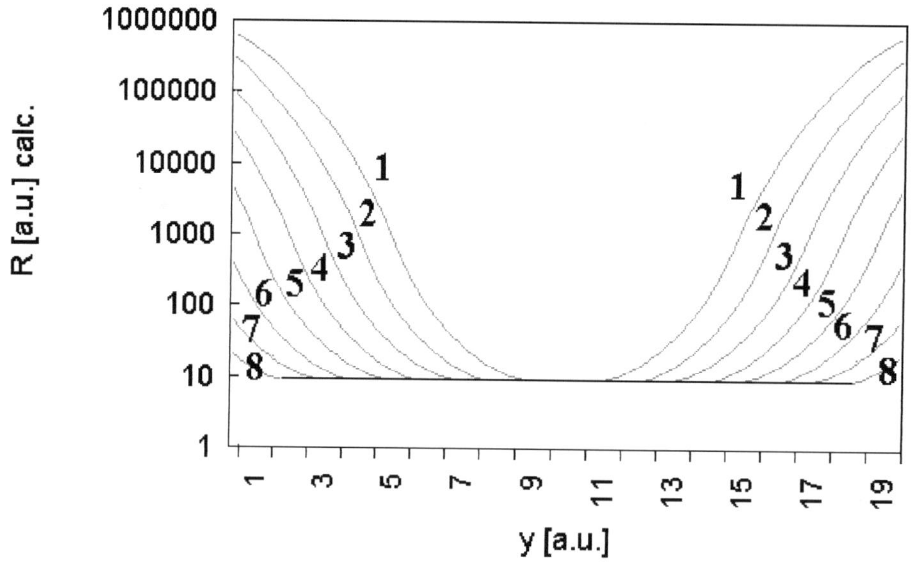

Fig. 9
Evolution of conductivity peak at a certain position following an additive behaviour of inverse local conductivities (resistance) during a RTD measurement using integrated microelectrodes for transduction

Application of Fundamental Measurements to Modeling and Process Development

Douglas Galloway, UOP LLC, Des Plaines, IL 60017

Commercial needs are consistently increasing the knowledge requirements for process development. As the pace of process development work increases, and the search for novel catalytic or adsorbent materials intensifies, an increased fundamental understanding of a given process is needed to gain further insight into process mechanisms and to improve testing capabilities. An integral part of addressing these needs is the development and application of advanced characterization tools. Using the proper characterization tools offers greater insight into identifying new adsorbent or catalytic materials, and also provides the fundamental parameters necessary for developing and optimizing a process. This talk will present several reactivity and adsorption characterization tools that are used at UOP during process development. The use of models necessary for both interpreting and summarizing process data will also be discussed.

Real Time Characterization of Micro Scale Processes

Brian Marquardt and Ray Chrisman. Center for Process Analytical Chemistry (CPAC) in the Chemistry Division, University of Washington, Seattle, WA 98195

As efforts to characterize reactions in micro scale equipment continue to grow new tools are being added to facilitate the operation. This presentation will demonstrate the use of some of these new tools such as NeSSI based tools and describe how they can be used to more rapidly and efficiently aid reaction characterization. The presentation will also demonstrate the real time characterization of several reaction types and show how the modular equipment enables new approaches to insure higher quality data gathering.

ON-LINE SAMPLING AND GAS CHROMATOGRAPHIC METHOD FOR A MICRO-SCALE GAS-LIQUID OXIDATION REACTOR

Patrick L. Mills
Department of Chemical and Natural Gas Engineering, Texas A&M University-Kingsville
700 University Blvd, MSC 188, Kingsville, TX 78363-8202 USA

Thomas M. Delaney and J. Scott McCracken
DuPont Company, Experimental Station
Rte 141 & Rising Sun Lane, Wilmington, DE 19880-0357 USA

Abstract

The development and application of a novel 5-cc micro reactor system for gas-liquid hydrocarbon oxidation kinetics studies under commercially-relevant conditions is described. A unique system feature is the use of a novel automated method for direct sampling and analysis of the liquid phase. This is accomplished by means of a custom-designed sampling valve that is integrated into the micro reactor body and connected to a GC system for direct injection of the vaporized liquid sample via heated capillary transfer tubing. The advantages of this approach over conventional manual sampling include: (1) the effect of sample size on the instantaneous reactor volume is negligible; (2) post-reaction of the sample once it exits the reactor is minimized; (3) the sample statistics are improved over conventional sampling methods; (4) light-boiling components and dissolved gases are directly captured for subsequent analysis; (5) reactor operation is more safe than commercially-available laboratory reactors owing to the small volume (5 cc or less); and (6) unattended operation with precise intermittent sampling of the liquid phase is possible.

The sampling concept was first demonstrated using a custom-designed test vessel whose geometry mimicked a 5 cc stirred gas-liquid microreactor. Two modes of liquid sampling were evaluated: (1) flashing of the liquid sample into a calibrated loop of a VICI® Valco six-port gas sampling valve; (2) flashing of the liquid sample directly into the injection port of a HP 5890 Series II gas chromatograph. A variation of the second mode involved the addition of a ten-foot heated section of capillary tubing between the exit of the vessel and the GC injector. The test samples consisted of an aqueous mixture of ca. 1 % each of C_1 to C_4 alcohols, and a crude reaction product obtained from a commercial-scale cyclohexane oxidation reactor.

It was found that the ability to fill the Valco valve sampling loop was dependent upon the pressure developed by the flashing liquid. For a fixed vessel pressure, the amount of vapor in the loop reaches a maximum and then decreases as the pulse valve duration increases. This occurs because the pressure created by the flashing liquid greatly reduces the amount of sample that can exit the valve. In the case of the direct injection methods, this effect was not observed. Hence, the total

GC area counts were reproducible within 0.1 % for a fixed value of the pulse valve duration. The variation that uses the transfer line is particularly useful for applications where the reactor must be located some distance from the GC system. Other details on the valve design and test results are also provided.

Introduction and Background

Selective oxidation processes have significant commercial importance in the chemical industry since they account for about 60% of the total annual production rate of chemical intermediates and final products [1]. These intermediates are often the monomers or key ingredients used to manufacture various polymers for a wide range of applications involving high performance plastics and textile fibers [2]. Oxidation reaction chemistries can be generally classified as either gas-liquid or gas-solid reactions that are catalyzed using either soluble organometallic complexes [3] or solid heterogeneous catalysts [4]. Process innovations are largely driven by the need to develop more attractive manufacturing economics [5-8] while reducing or eliminating emissions for improved environmental impact [9]. A review of catalytic technologies that have been commercialized in the USA during the 1990's demonstrates that partial oxidation catalytic technologies continue to occupy a position of importance based upon the number of new patents that appear and new processes that are under development [10].

The liquid-phase oxidation of cyclohexane to cyclohexanone (K), cyclohexanol (A), and cyclohexylperoxide (P) [11] exemplifies the scope of gas-liquid oxidation chemistries since the reaction products are key monomers used in the manufacture of the raw materials for nylon polymers and textile products, such as caprolactam [12] and adipic acid [13]. The first group of commercial processes for cyclohexane oxidation are based upon the use of soluble cobalt or other metal salts to catalyze the reaction to produce a mixture of so-called "K/A" oil. The second group of processes are essentially uncatalyzed since they convert cyclohexane to the cyclohexylperoxide intermediate "P" followed by decomposition to K/A oil [14]. Although details of the reactor designs used in these commercial processes are kept as proprietary, it can be surmised that they utilize a single or series of continuous-flow stirred tanks or bubble column reactors where the latter may incorporate various internals to alter the gas and liquid flow patterns. No matter which reactor type is used, operation is constrained by the need to maintain reaction mixtures within either the lower or upper flammability limit to prevent a deflagration.

In the mid-1990's, Praxair, Inc issued various patents (see [15] for an example) that disclosed the use of pure oxygen versus air or enriched air in an effort to develop safer and more efficient processes that employ gas-liquid oxidation reactions. This technology is based upon a special-purpose reactor design called a Liquid Oxidation Reactor (LOR) as an alternate approach versus conventional stirred tanks and bubble columns. The LOR represents a form of process intensification and sustainable development for the following reasons: (1) the amount of process vent gas that must be treated is reduced when compared to air-based processes since inert N_2 is present at low concentration levels; (2) due to a higher concentration of dissolved di-oxygen in the liquid-phase, the reaction can be operated at lower temperatures and/or pressures versus conventional processes so that higher product selectivity is obtained; and (3) the increased utilization of oxygen leads to more efficient consumption of raw materials with lower selectivities to undesired reaction by-products and reduced environmental impact. A good discussion of this particular technology that analyzes safety aspects for cyclohexane oxidation processes is provided in a recent review [16]. Despite the apparent advantages of the LOR technology versus conventional oxidation reactors, no commercial plant that uses this design has been implemented, at least to the authors' knowledge.

Detailed reaction engineering analysis for either conventional gas-liquid oxidation reactors or the LOR design as reported in the open literature is complicated by the lack of realistic intrinsic kinetic models. This lack of robust kinetic models for either catalyzed or uncatalyzed operation has most likely limited advances in the development and implementation of improved and new processes. The key literature and status of kinetics models has been summarized in recent work concerned with the interaction of bubble dynamics and kinetic-transport limitations [17]. A review of the works cited therein and other recent publications on cyclohexane oxidation kinetics suggests that a opportunity exists to fill an existing gap on both fundamental and more applied aspects for this and other commercially-important gas-liquid oxidation reactions, such as the oxidation of *para*-xylene to terephthalic acid [18]. Furthermore, patent reports teach that uncatalyzed operation has a number of process advantages when compared to catalyzed operation [14]. Consequently, the ability to predict the preferred process configuration and process conditions for the uncatalyzed versus catalyzed mode of operation is not possible using first principles until detailed kinetic models are developed and incorporated into the requisite reactor engineering models .

Unfortunately, the study of gas-liquid oxidation reactions in the presence or absence of a catalyst generates a variety of outstanding experimental challenges that are either absent or of lesser concern when compared to other classes of gas-liquid catalyzed systems, such as hydrogenations, hydroformylations, carbonylations, chlorinations, fluorinations, and nitrations. Some of these experimental challenges include: (1) the presence of various explosion hazards that can occur when blending hydrocarbons with either air, pure oxygen, or enriched air; (2) the effect of various reactor materials of construction and other reactor system wetted parts on the observed reaction chemistry; (3) sampling and analysis of the primary and secondary reaction products, such as organic alcohols, aldehydes, carboxylic acids, and other oxygenated species, so that the resulting compositions represents those which exist in the hot reaction environment; (4) variation of the oxygen partial pressure over a sufficiently broad range that its effect on the reaction kinetics (rates, conversion, selectivity, etc.) can be properly discriminated and optimized; and (5) controlling reaction light-off and reaction extinction so that the reaction can be safely controlled and studied. A discussion on these and other related issues is not provided here for brevity since these will be analyzed in more detail in a future publication [19].

One objective of this paper is to describe a micro-scale laboratory reactor for gas-liquid oxidation chemistries that can be used for a range of experimental studies, such as overall reaction performance or for measurement of detailed kinetic data for either catalyzed or uncatalyzed modes of operation. The reactor vessel is selected so the total liquid volume is on the order of 5-cc or less. This feature satisfies process hazards requirements from an equivalent TNT (trinitrotoluene) perspective for safe operation in a standard chemical fume hood with proper deflection shielding. Of particular interest in this work is the development of a system for *in-situ* automated sampling of this particular micro-scale gas-liquid reactor design that overcomes many of the disadvantages associated with classical modes of sampling, such as multi-port sampling valves. The ability to perform on-line sampling and analysis the reaction mixture in gas-liquid lab reactors with liquid volumes on the order of a few cc is one of the more difficult problems in microreaction technology. For this reason, emphasis was placed upon developing a novel integrated sampling and analytical system for this application.

Experimental

1. Sampling Valve Test Apparatus

A diagram of the prototype test apparatus that was designed and constructed to evaluate the sampling-injection system is shown in Figure 1. This system was not intended for hot reaction studies with liquid hydrocarbons and oxygen-containing gases, but to provide an experimental platform for development and testing of the integrated sampling-injection valve and on-line GC analytical system. The vessel was constructed of aluminum metal for ease of machining, and had an inside height of 1.25 inches with an inside diameter of 0.5625 inches, which translates into a working volume of about 5-cc. Heating was accomplished using six 25-watt cartridge heaters. Four of the heaters were equidistantly placed in the vessel wall, while the remaining two were located in an aluminum shell clamped to the sample injection valve. Temperature control for both the vessel and injection valve was provided using two independent Eurotherm PID controllers. The base of the vessel was designed so the sampling valve body could be screwed into position and sealed with a metal gasket, such as thin gold foil. The base contained an orifice with an opening of 0.020 inches. Other openings (*e.g.*, 0.010 inches to 0.040 inches) could be used, but this particular one was determined to allow a small liquid sample (on the order of a few micro liters) to pass into the sample valve flash chamber. The injection valve used in the experiments described here was identical to the pulse valve design used in the DuPont TAP (Temporal Analysis of Products) reactor valve manifold. The poppet was constructed of Vespel polymer and was a custom design. The top of the vessel contained one port for attachment of a pressure gauge and a second port for purging and pressurization with inert nitrogen gas. Later modifications included: (1) the addition of a port for purging of the direct injection valve with helium carrier gas, or purging of the combined injection-Valco valve (described below) with nitrogen; and (2) modification of the injection valve internals to decrease the internal volume.

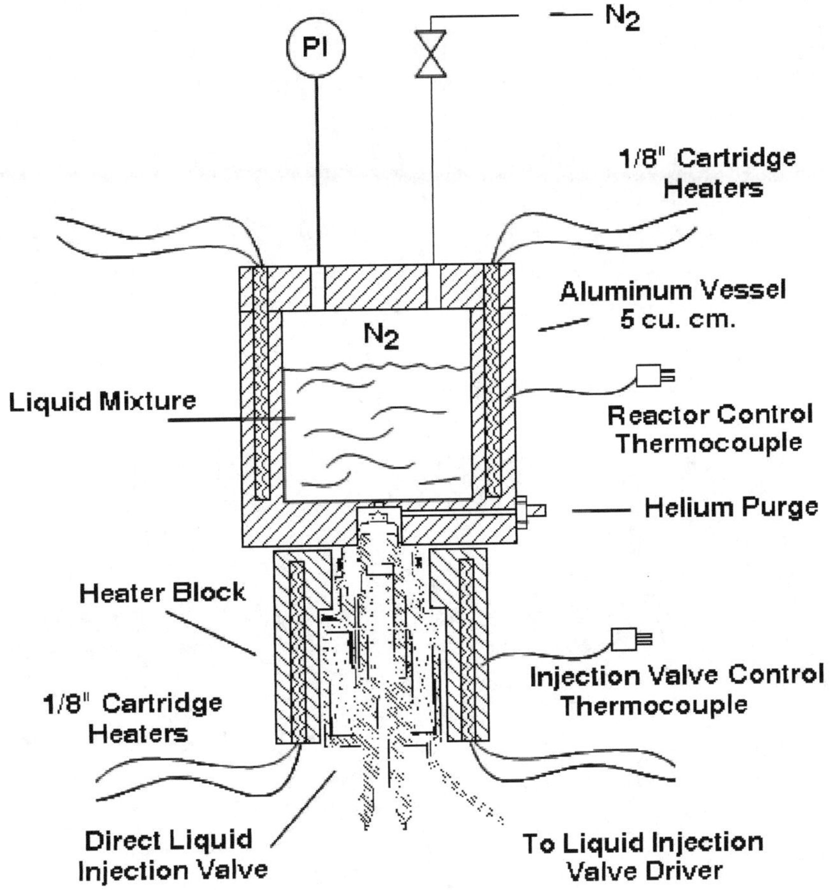

Figure 1. Combined micro reaction vessel and direct injection valve.

2. *Combined Microreactor-Sampling Valve Assembly*

Once the preferred operating mode for sampling the liquid-phase using the above test apparatus was determined, a prototype micro-scale gas-liquid reactor was designed and fabricated. A photograph that shows the key components of this design is shown in Figure 2. To provide a perspective of size, a United State Treasury coin (quarter) is also shown in the top section of the figure. The reactor in the top part of the figure was machined from a solid piece of type 316 stainless steel cylindrical rod. The bottom (right-side in the figure) was drilled to accept the male threads of the sampling valve, while the top (left-side in the figure) was drilled-through and contained a knife edge flange so that the reactor top could be sealed with a metal. The vessel had an inside height of 1.25 inches with an inside diameter of 0.5625 inches. These dimensions were selected to match those for the aluminum test vessel shown above in Figure 1. Mixing of the reactor contents was accomplished using a 4-bladed stirrer-shaft assembly that was fabricated so it was held in place by a self-supporting 4-baffle assembly. The outer dimensions of the baffle assembly were selected so the baffles formed a friction-fit inside the reactor body. The upper part of the shaft contained a rare earth magnet whose field was strong enough to adequately couple to a magnet of the same type, but which was located on a second shaft located in close proximity of the reactor top. This second shaft was supported by an external bearing assembly and was driven by a small DC motor with a variable speed controller. Reactor heating was accomplished by placing the assembly inside a heated convection oven that was purged with inert nitrogen. The reactor mass was large enough that isothermal operation was achieved. The upper temperature limit (250°C) was defined by the materials used to fabricate the electromagnetically-driven coil shown in Figure 2 that is a key component of the injection valve. Reaction gas was added upon demand from a small reservoir that was operated at a constant downstream pressure as measured in the reactor head space. The gas was introduced through a port located on the bottom of the reactor. The acrylic version shown below the metal reactor assembly was used for flow visualization studies. Additional details are provided elsewhere [19].

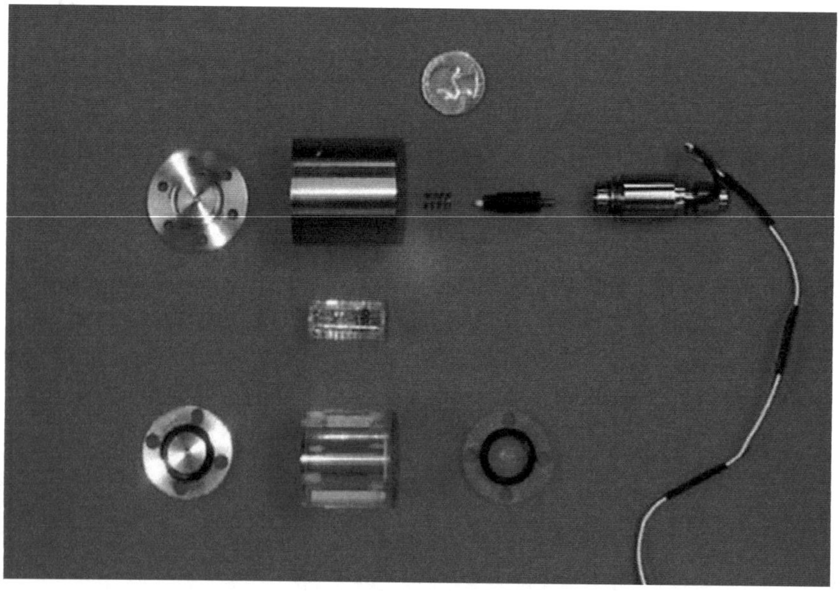

Figure 2. Photograph of the prototype micro-scale reactor showing the key components of the reactor body and sampling valve. The acrylic version shown at the bottom was used for flow visualization of gas-liquid mixing.

4. *Operating Modes*

Capture and injection of the liquid sample from the reaction vessel was performed using several modes as part of an effort to identify the preferred approach. These modes included a *direct injection mode* and a *combined injection-Valco valve mode*. In the so-called *direct injection mode* (not shown), the exit of the direct injection valve shown in Figure 1 was directly connected to the GC injection port using a short section (ca. 1.5 inches) of 1/16 inch o.d. type 316 stainless steel capillary tubing having an inside diameter of 0.04 inches. The septum nut was drilled out so that the transfer line extended into the injection port the same distance as a needle used for typical manual injection. The tubing was welded to the nut to achieve proper alignment in the injection port and also to prevent leakage of the carrier gas past the tubing. This method does not allow manual injections to be performed using a syringe. A variant of this mode used the helium carrier gas to transport the flashed sample through a ten-foot section of heated 1/16 o.d. capillary tubing to simulate remote sampling. This permitted manual injections with a syringe, since the tubing was connected to the carrier gas return line of the GC injector.

The *combined injection-Valco valve mode* utilized a six-port Valco valve for capture of the flashed liquid. This experimental arrangement is illustrated in Figure 3. Here, the exit of the liquid injection valve is connected to a six-port Valco gas-sampling valve. By opening the liquid injection valve poppet for a fixed duration, the liquid is flashed into the loop of the Valco gas sampling valve. Unlike the direct injection mode, this allows, in principle, a known amount of vapor to be captured and sent to the GC for subsequent analysis. This would be useful, for example, in applications where knowledge of the sample amount was needed for quantification or where multiple reactors would be connected to a single GC.

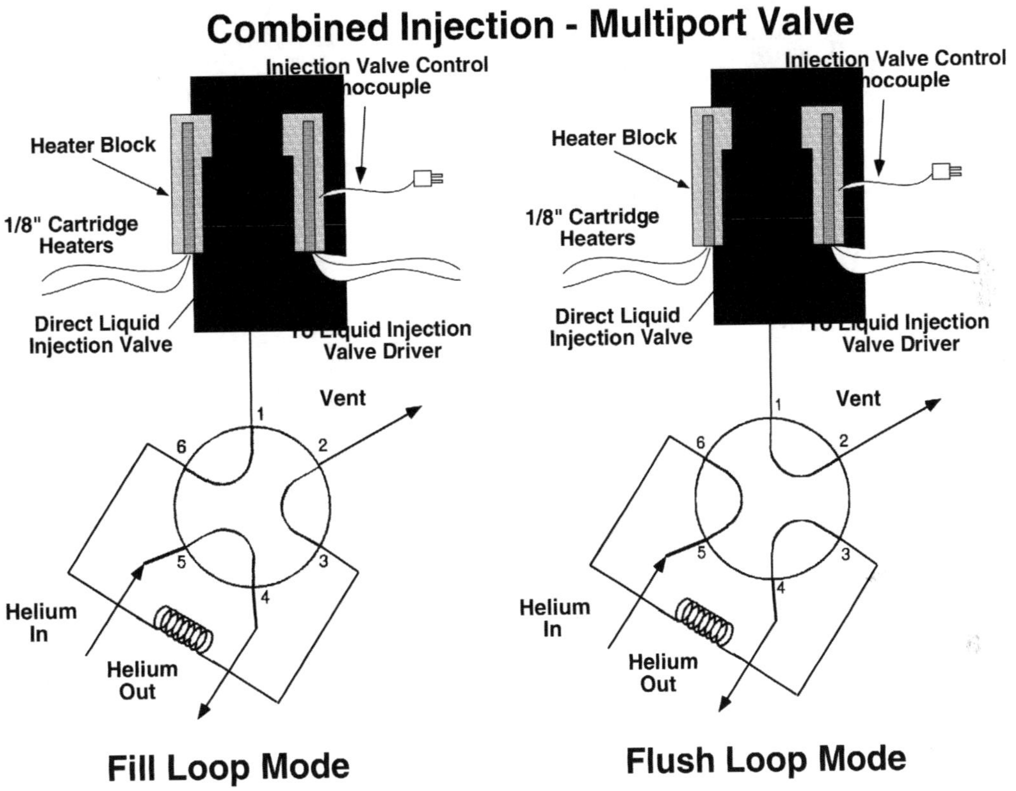

Figure 3. Combined liquid injection-Valco valve mode.

3. *Analytical*

a. Equipment. The analyses were performed using a Hewlett-Packard (Agilent) Model 5890 Series II gas chromatograph equipped with both a flame ionization detector (FID) and a thermal conductivity detector (TCD). It was also equipped with a Series 5971 Mass Selective Detector (MSD) for positive identification of the eluting peaks. The FID side was used for analysis of the hydrocarbons and oxygenates using capillary columns operated in the split mode with a 40:1 split ratio, and was equipped with a split/splitless capillary injector. The TCD was not used.

Helium was used as the capillary column carrier gas at volumetric flow rates of 10 cm^3/min. Make-up gas was added before the FID to achieve a total flow rate of 30

cm^3/min to the detector. The hydrogen and air gas flow rates to the FID were set at 30 cm^3/min and 450 cm^3/min, respectively, as specified by the manufacturer. The temperatures for the FID side injector and detector were set at 230°C and 250°C, respectively. The FID electrometer sensitivity was usually set at a range of six and an attenuation of four.

The analog output signal from the FID was connected to a HP Chemstation data analysis system. The initial system software, which was version V.02.02, was loaded on a HP Vectra 386/25 IBM compatible computer. It was used for all processing and analysis of the raw FID and mass spectrometer output, graphical display on a CRT, and storage on the hard disc. The output signal for the FID was also connected to a Hewlett-Packard Model 3396 Integrator and used for local display on chart paper.

Response factors for the test mixtures and anticipated primary and secondary products were determined using manual injection with an injection volume of 0.5 μl. The area counts for each identified species were taken from either the HP 3396 integrator or HP Chemstation and processed using a Microsoft Excel program.

b. Columns. A variety of capillary columns with a wide range of polarities were evaluated for peak resolution and retention time using various liquid and gas test mixtures (*vide infra*). The compositions of these test mixtures were based upon previous catalyst testing experience as well as reaction product mixtures reported in the open literature using similar catalysts. A summary of the columns and their specifications is given in Table I. The columns were used as received from the commercial supplier, and were conditioned prior to conducting any test evaluations following the manufacturers' recommendation.

Selection of the capillary columns listed in Table I for analysis of the hydrocarbon and oxygenated reaction products was performed by considering the effect of stationary phase polarity on selectivity. Polarity generally indicates the way a given solute interacts with the stationary phase, and is quantified using numerical values of the McReynolds constant. The latter quantity is assigned a value of zero for a nonpolar phase, such as DB-1, while values between 2000 to 3000 are indicative of polar stationary phases, such as DB-FFAP. Separation according to boiling point is the basis for selection of nonpolar phases, while stationary phases having intermediate polarity, such as 50% phenyl- and 50% methylpolysiloxane or DB-17, retain species based upon both boiling point and induced dipole interaction or hydrogen bonding. Polar and strongly polar phases, such as TPA modified polyethylene glycol or DB-FFAP, retain polar compounds longer than nonpolar compounds, due to the induced dipole-dipole interactions between the functional groups of the same sample and stationary phase. For the cyclohexane oxidation reaction mixtures, all columns were evaluated for peak resolution and to test for reversible absorption of the various species.

Table I. Columns Evaluated for Analysis of C_1 to C_6 Hydrocarbons and Oxygenates

Column Name	Column Type	Column Dimensions: Diameter	Length	Film Thickness
J&W DB-1	megabore	0.53 mm	30 m	5.0 µm
Restek 1701	megabore	0.53 mm	30 m	3.0 µm
J&W DB-FFAP	megabore	0.53 mm	30 m	1.0 µm

c. Test Mixtures and Calibration Standards. Several liquid mixtures as well as various pure components in these mixtures were used to evaluate the columns listed above in Table I. The pure gas and liquid components were obtained from commercial sources and used as received without any further pre-treatment. A gas mixture calibration standard containing various C_1 to C_4 hydrocarbons in N_2 was also obtained from a commercial source (Roberts Oxygen, West Chester, PA).

Various pure organic species and their absolute weights that were used to prepare four different liquid standards. Standard one contained nearly all of the important C_1 to C_6 alkanes, olefins, acids, aldehydes and ketones that have been reported as primary or secondary products in cyclohexane partial oxidation (see relevant references cited in [17]). Standard two was the same as standard one, except methyl ethyl ketone and glyoxalic acid were omitted and 2,3-dihydrofuran was included. The former two components were omitted since they either did not elute, or mass spectrometry analysis showed that they co-eluted with other more important species. Standard three was the same as standard two, except methacrolein was omitted. Mass spectrometry analysis (*vide infra*) showed that methacrolein contained a pyran derivative impurity that co-eluted with either acrylic acid on the DB-1, or maleic anhydride on the 1701 column. Standard four contained furan and selected derivatives, along with acetic acid and acrylic acid in acetone solvent. This standard was intended to provide an independent confirmation of the response factors for comparison to those derived using the previous three standards.

The above-described mixtures were prepared by standard gravimetric techniques using a pan balance accurate to within +/- 0.0001 grams with a precision of +/-0.005 grams. Acetone was selected as a solvent due to solubility considerations. To minimize errors associated with weight losses due to volatility, the solvent and the lower boiling liquid components were cooled in dry ice or maintained in a refrigerator prior to weighing. These same mixtures were also used to evaluate both absolute and relative response factors. In some cases, manual injections of pure compounds or binary mixtures were used to determine either absolute or relative retention times, and to acquire mass spectral fragmentation patterns for addition to the HP Chemstation library.

Results and Discussion

The results of various experiments using both the direct injection and combined direct injection-Valco valve modes are summarized below.

1. Direct Injection Valve Mode

Figure 4 shows the percentage peak area for methanol, ethanol, and propanol as a function of the direct injection valve opening duration where the latter has units of microseconds. In this case, the sample was flashed directly into the GC injection port through the short section of heated transfer line as described above. The data show that above valve durations of 500 μs, the percentage peak areas for all components are constant. Hence, increasing the total amount of sample introduced results in a uniform vaporization of the individual components that are present in the mixture. If the liquid vaporization was non-uniform, then the percentage peak areas for each component would not be linear with increasing duration, since the latter is proportional to sample size. The choice of direct injection valve duration time for a particular application would depend upon such factors as concentration of components in the mixture, capillary column split ratio, and other parameters that affect the column loading. For the current application the indicated times were sufficient.

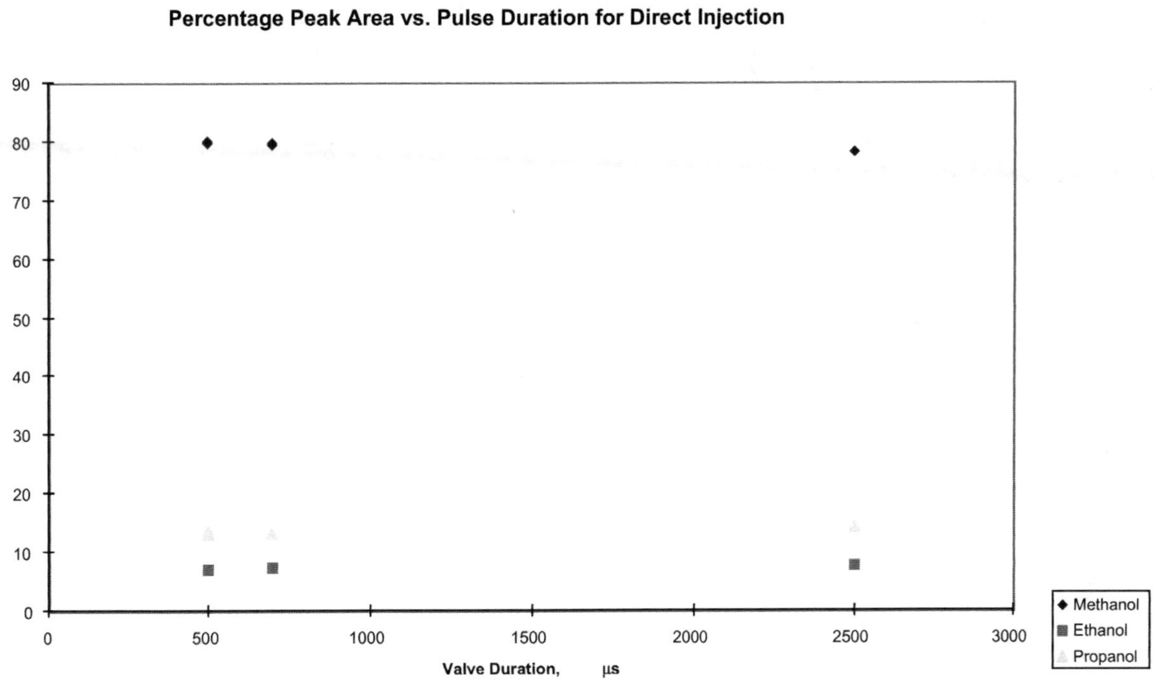

Figure 4. Percentage peak area for various alcohols in the liquid test mixture as a function of the direct injection valve duration.

Figure 5 shows the total peak area of the test mixture as a function of the direct injection valve duration. The total peak area shows a slight nonlinear trend with respect to the valve duration. It was postulated that the amount of sample introduced is not linearly related to the valve opening time. However, the reproducibility of the peak areas at any given duration is excellent, since the standard deviation of the error is within the width of the data points.

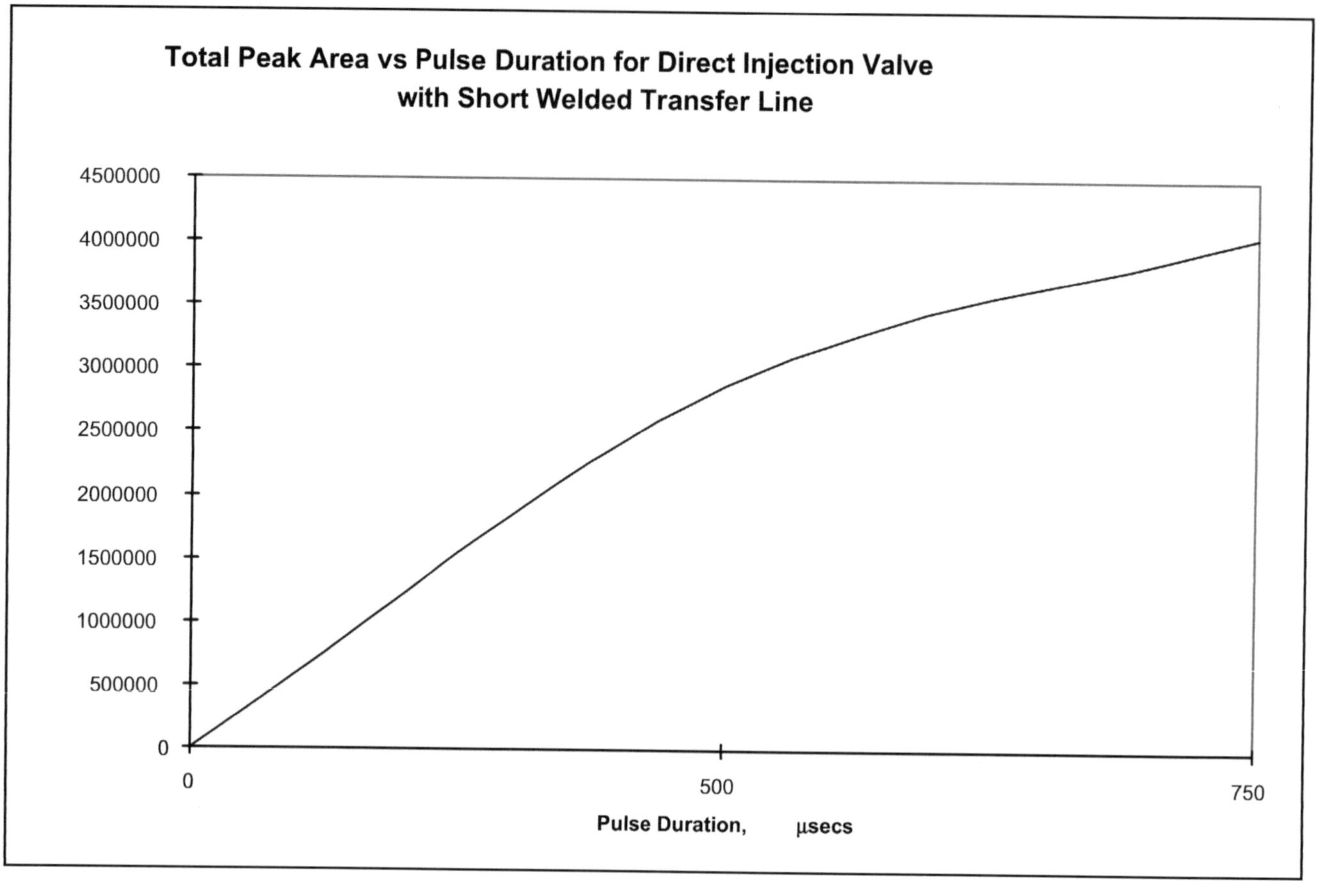

Figure 5. Total peak area for various alcohols in the liquid test mixture as a function of the direct injection valve duration.

Figure 6 shows the results obtained when the system was operated in the direct injection mode, except that a ten-foot section of transfer line was used. Here, the total peak area is nearly linear with respect to the valve injection duration. The reproducibility of the injection process is again excellent, since the indicated data points are based upon the average of five injections whose standard deviation is within the width of the data points. This mode would be very useful for applications where the reactor or process to be sampled is located remotely from the GC analytical system where direct injection into the GC injector is not practical or desirable.

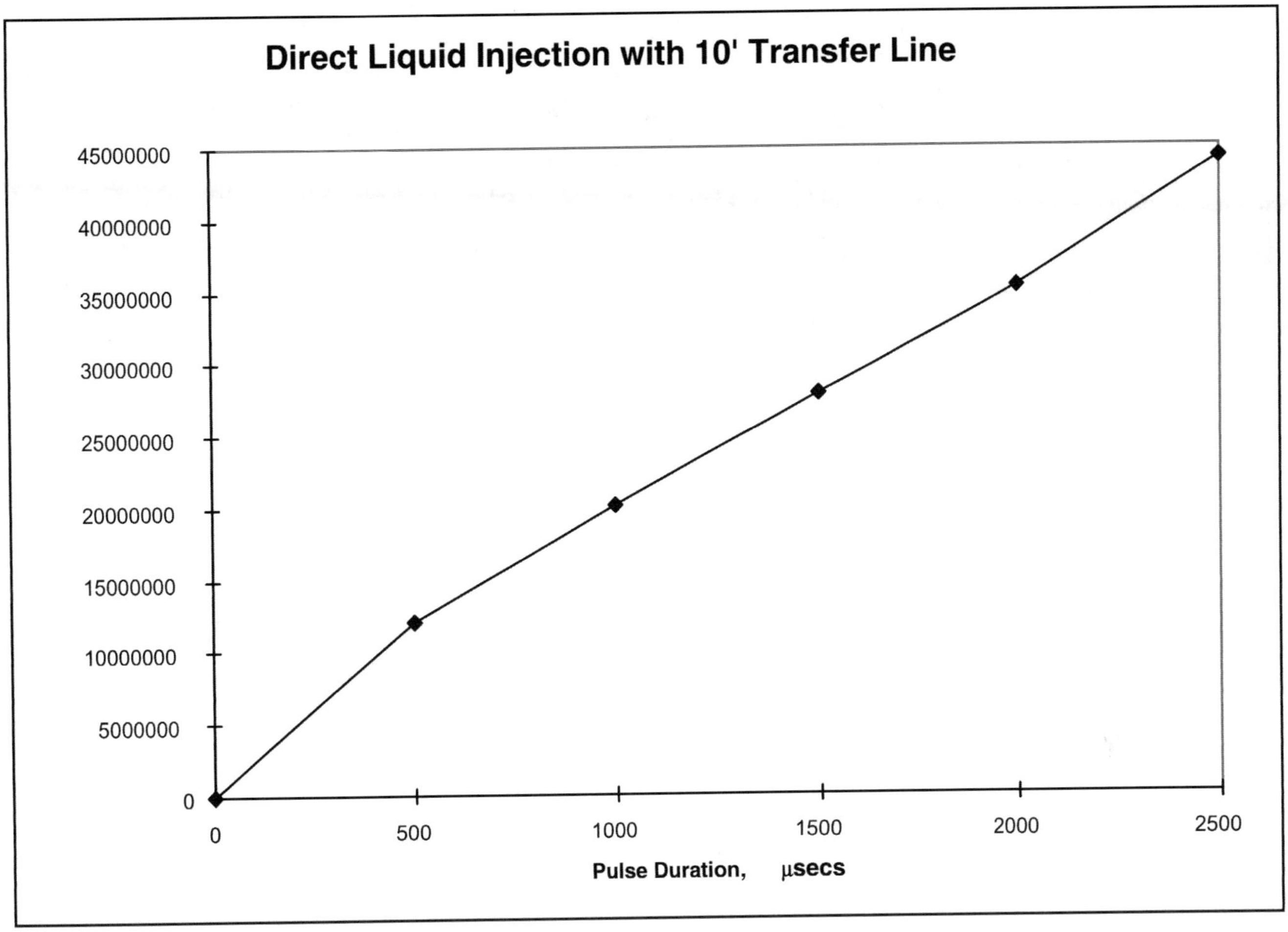

Figure 6. Total peak area versus valve opening duration using direct liquid injection with a ten-foot section of transfer line.

2. Combined Direct Injection-Valco Valve Mode

Figure 7 shows the total peak area versus valve duration for the combined direct injection-Valco valve mode using a constant vessel head pressure of 200 psig. The total peak area increases, reaches a maximum, and then decreases with increasing values of the valve duration. When a liquid flashes into a constant volume, the total pressure will increase as a result of the vaporization process. Hence, the pressure generated by the flash creates a back-pressure on the vapor side of the direct injection valve. The resulting pressure difference between the vessel (process side) and the vapor space (sample valve side) decreases the driving force for injection of the liquid. Eventually, the pressure generated by the flashing liquid sample is large enough that the amount of vapor sample generated decreases with further increases in the valve duration. This is shown in Figure 7 where the amount of sample in the Valco valve loop decreases once the valve duration is increased beyond a duration of ca. 500 µs. This is the main reason that modern GC injector inlets which use capillary columns employ a septum purge. The purge reduces the amount of vapor generated after the sample is introduced and vaporized so that sample may be carried into the column inlet.

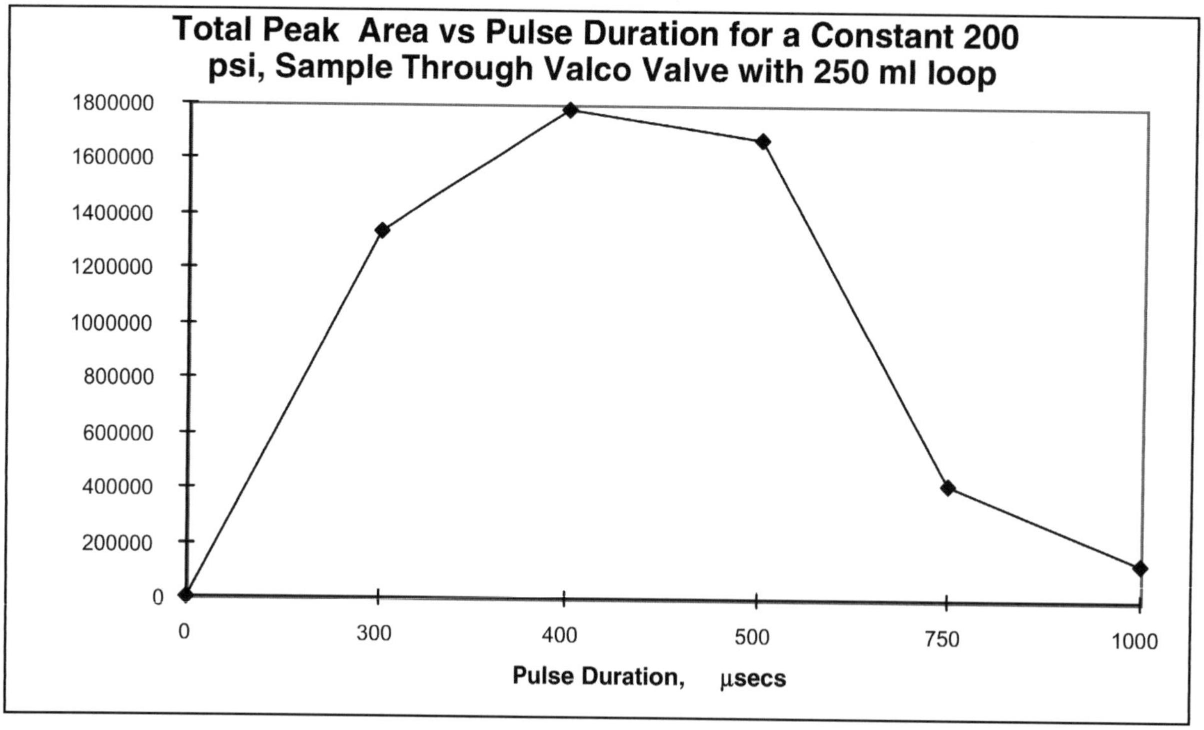

Figure 7. Total peak area for various alcohols in the liquid test mixture as a function of the direct injection valve duration using the combined system.

3. Reaction Mixture Results

Figure 8 shows the chromatogram obtained when a synthetic mixture containing ca. 1 wt % each of valeric acid, cyclohexanol, cyclohexanone, caprolactone, hexanoic acid, CHHP, valerolactone, chlorobenzene (internal standard) in cyclohexane are analyzed using a nonpolar DB-1 column. The chromatographic conditions are given in the figure caption. The sample was first analyzed by manual injection, and then repeated by charging the test apparatus shown in Figure 1 with a few cc of the mixture. For the latter case, vessel was heated to 100°C and pressurized with nitrogen to 250 psig of head pressure. The GC helium carrier was allowed to flow continuously through the sample injection valve chamber. Injection was accomplished by pulsing the valve for a duration of 100 µs so the sample was flashed into the carrier stream. Analysis of the mass spectrum for each peak shows that 1,1-oxybis ethane is present as an impurity in one of the added components. Both the chlorobenzene-valeric acid and cyclohexanol-cyclohexanone pairs are not baseline resolved using the indicated chromatographic conditions, although the apparent quality of the peaks appears to be satisfactory for these initial trials. The chromatograms obtained by manual injection versus the direct injection valve could not be distinguished when they were directly overlapped, which demonstrates operability of the proposed technique.

Figure 9 shows the chromatogram obtained when a reaction product from a commercial cyclohexane oxidation reactor was analyzed using the same chromatographic conditions given above in Figure 8 with chlorobenzene added as an internal standard. The vessel was heated to 100°C and pressurized with nitrogen to 250 psig of reactor head pressure. Identical results were obtained by both manual injection and the direct injection sample valve, which is parallel to those for the test mixture described above. As expected, the chromatogram is more complex and shows that a wide spectrum of hydrocarbons and oxygenated hydrocarbons are obtained in the C_1 to C_6 range. Cyclohexane gives an asymmetrical peak and contains a number of species that elute on the tailing portion. Some of the species that are present in small concentrations were not identified. As shown above in the synthetic sample, cyclohexanol and cyclohexanone are not baseline resolved so that other conditions or another stationary phase appear to be necessary. Information on the primary and secondary products gleaned from this type of analysis will be useful in defining the reaction pathways as part of the kinetic analysis. This is a topic for future study.

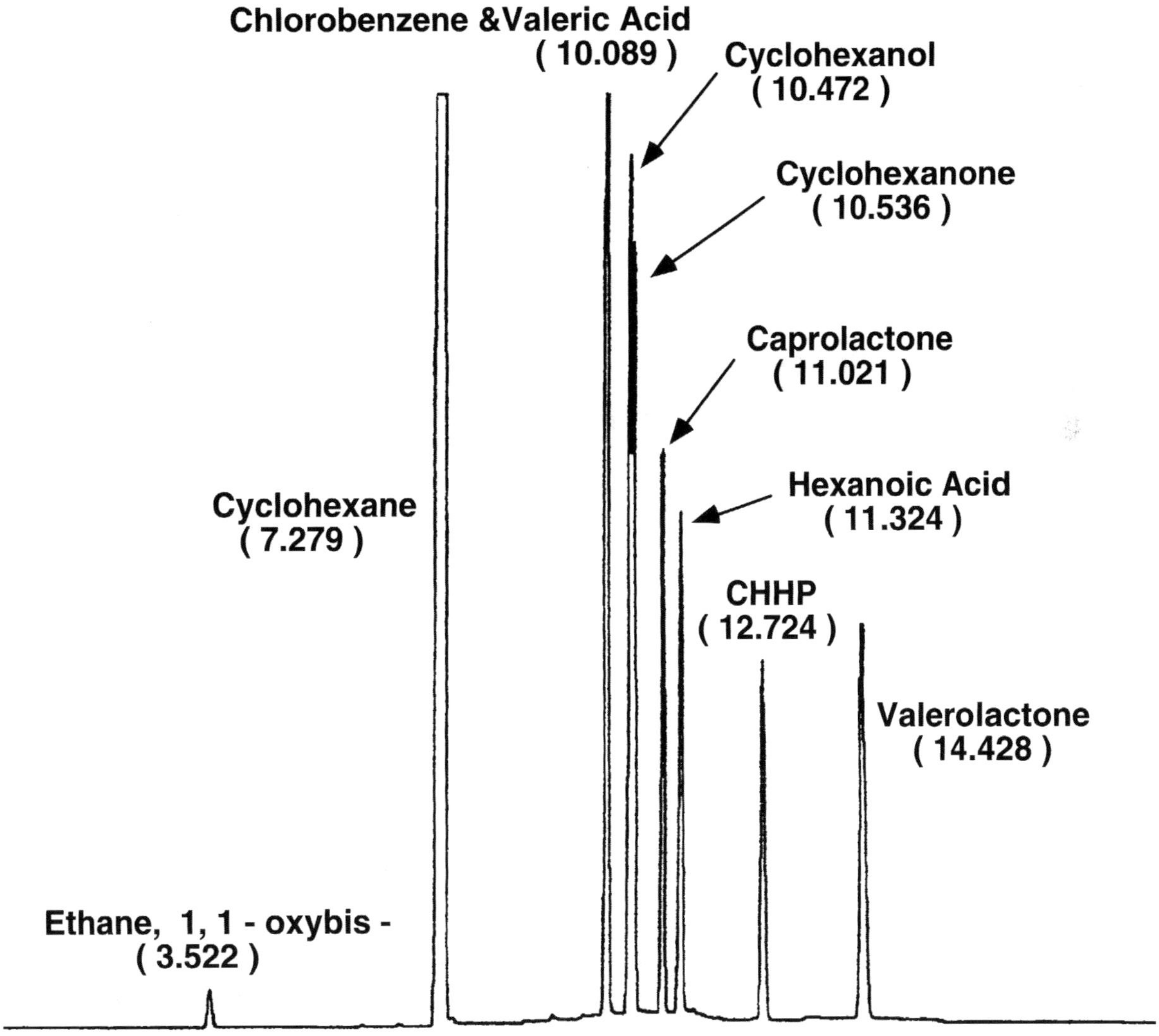

Figure 8. Chromatogram of a synthetic cyane oxidation reaction mixture using a nonpolar DB-1 capillary on the FID. Conditions: 35 °C for 12 mins., then 20 °C/min to 165 °C; 0.5 µl manual injection, range = 3, split ratio = 40:1.

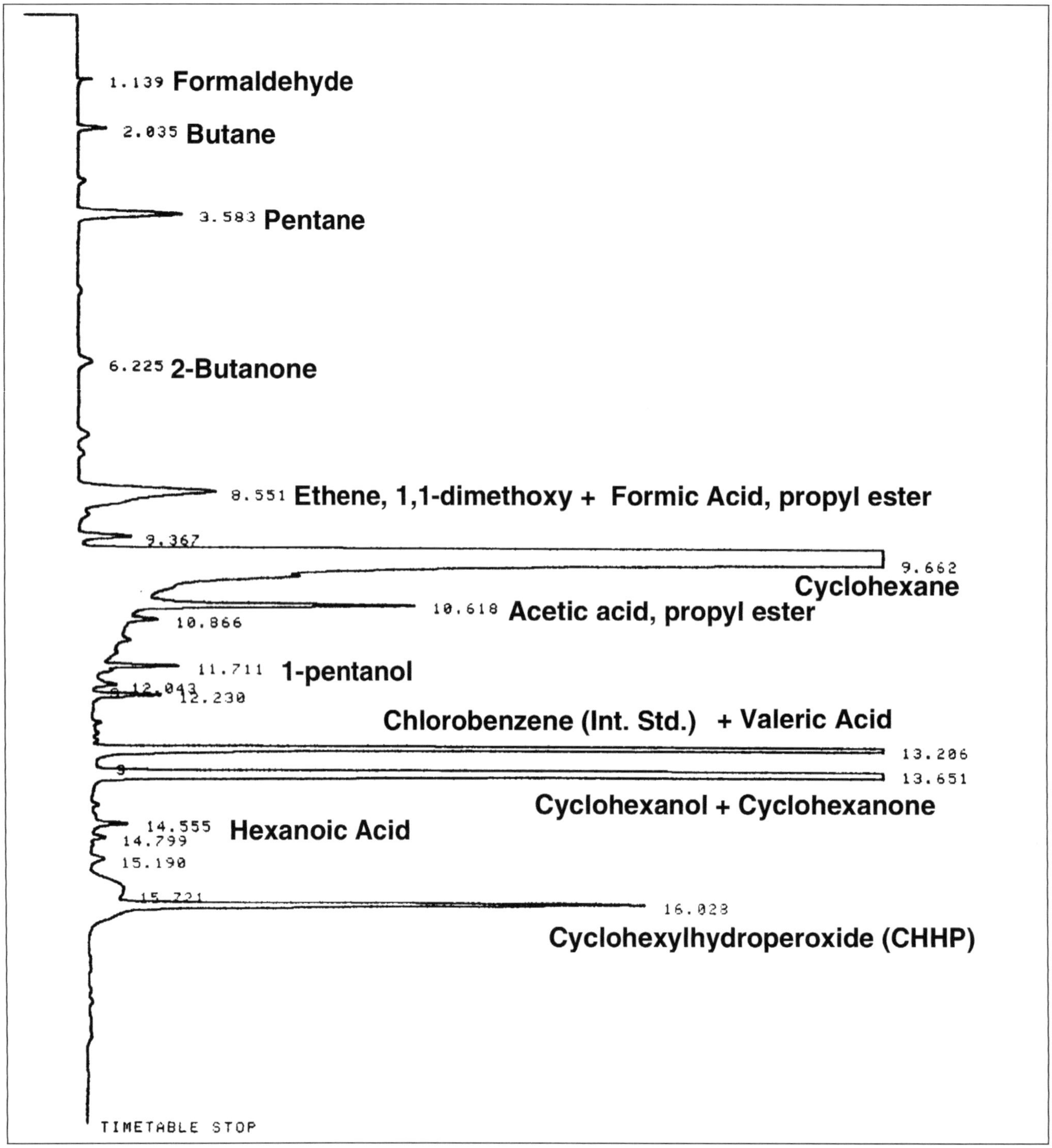

Figure 9. Chromatogram of a cyclohexane oxidation sample from a commercial reactor using a nonpolar DB-1 capillary on the FID. Conditions: 35 °C for 12 mins., then 20 °C/min to 165 °C;
0.1 μl manual injection, range = 3, split ratio = 40:1.

Conclusion and Future Efforts

A novel method for sampling the liquid phase in laboratory-scale gas-liquid microreactors has been developed. Although it was originally intended for on-line, unattended operation of these systems, it could conceivably be applied to any micro system where the sampling and analysis of liquids using on-line GC or HPLC methods is desired. Several options with regard to implementation of the methodology have been identified, that include: (1) direct injection using either short or extended sample transfer lines, or (2) one that combines the direct injection valve with a Valco gas sampling valve.

The vessel shown above in Figure 1 has been subsequently modified to include a set of baffles and internal stirring mechanism. The latter is based upon a magnetically coupled micro drive that uses rare-earth magnets. The demonstration of this system for collection of kinetic type data is a work in progress.

References

1. Centi, G.; Cavani, F.; Trifiro, F. *Selective Oxidation by Heterogeneous Catalysis*, 1st ed. Fundamental and Applied Catalysis, ed. M.V. Twigg and M.S. Spencer. Vol. 7. **2001**, New York: Kluwer Academic/Plenum. 505.

2. Ulrich, H. *Introduction to Industrial Polymers*. 2nd ed. Society of Plastics Engineers, ed. R.J. Ehrig. Vol. 18. **1993**, Munich: Hanser Publishers. 188.

3. Cornils, B.; Herrmann, W.A. (eds.) *Applied Homogeneous Catalysis with OrganometallicCompounds: A Comprehensive Handbook in Three Volumes*. 2nd ed. Vol. 1. **2002**, Wiley-VCH: New York. 1450.

4. Hodnett, B.K. *Heterogeneous Catalytic Oxidation: Fundamental and Technological Aspects of the Selective and Total Oxidation of Organic Compounds*. 1st ed. Vol. 1. **2000**, New York: Wiley-VCH. 360.

5. Arntz, D. *Trends in the Chemical Industry*. Catal. Today **1993** *18*, 173.

6. Cavani, F.; Trifiro, F. *Some Innovative Aspects in the Production of Monomers via Catalyzed Oxidation Processes*. Appl. Catal. A: General **1992** *88*, 115.

7. Holderich, W.F. *Catalyzed Processes - A Trend Analysis.* Appl. Catal. A: General **1994 113** 115.

8. J.J. Lerou, M.P. Harold, and P.L. Mills, *Industrial Heterogeneous Gas-Phase Oxidation Processes*, in *Catalytic Oxidation: Principles and Applications: A Course of the Netherlands Institute for Catalysis Research, 6-9 June,* **1994**, S.R. A. and R.A. van Santen, Editors. **1999**, World Scientific Publishing: Rolduc, Kerkrade, The Netherlands. p. 392.

9. J.N. Armor, *Striving for catalytically green processes in the 21st century.* Applied Catalysis A: General, **1999** 189(2): p. 153-162.

10. J.N. Armor, *New catalytic technology commercialized in the USA during the 1990s.* Applied Catalysis A: General, **2001** 222(1-2): p. 407-426.

11. W. B. Fisher and J. F. vanPeppen, *Cyclohexanol and Cyclohexanone*, in Kirk-Othmer Encyclopedia of Chemical Technology, Article On-line Posting Date: December 4, 2000.

12. W. B. Fisher and L. Crescentini, *Caprolactam*, in Kirk-Othmer Encyclopedia of Chemical Technology, Article On-line Posting Date: December 4, 2000.

13. Adipic Acid, in Kirk-Othmer Encyclopedia of Chemical Technology Article On-line Posting Date: December 4, 2000.

14. H. A. C. Baur and U. F. Kragten, *Process for the Preparation of a Cyclohexanone and/or Cyclohexanol.* European Patent No. 0367326 A1, May 9, 1990.

15. A. K. Roby and J. P. Kingsley (inventors), *Enhanced Oxidation of Organic Chemicals.* US Patent 5,536,875 assigned to Praxair Technology, Inc., July 16, 1996.

16. J-R Chen, *An Inherently Safer Process of Cyclojexane Oxidation Using Pure Oxygen-An Example of How Better Process Safety Leads to Better Productivity.* Process Safety Process. **2004**, 23(1) 72-81.

17. J. G. Khinast, A. A. Koynov, and T. M. Leib, *Reactive Mass Transfer at Gas-Liquid Interfaces: Impact of Micro-scale Fluid Dynamics on Yield and Selectivity of Liquid-Phase Cyclohexane Oxidation*, Chem. Eng. Sci. **2003** 58, 3961-3891.

18. Yan Xuefeng, *Data Mining Macrokinetic Approach Based on ANN and Its Application to Model Industrial Oxidation of* p-Xylene *to Terephthalic Acid*, Chem. Eng. Sci. (in press) *Available online 23 February 2007.*

19. P. L. Mills, *Laboratory-scale Reactors for Gas-Liquid Reactions – A Review*. Ind. Eng. Chem. Res., in preparation, **2007.**

Sensor Testing System for Environmental and Process Applications

Mark M. Disko[1], Andrew E. Riley[1], Cornelius H. Brons[1], Charles Rebick[1], Manuel A. Francisco[1], Boris Mizaikoff[2], and Christina Young[2]. (1) Corporate Strategic Research, ExxonMobil Research and Engineering Company, 1545 Route 22 East, Annandale, NJ 08801, (2) School of Chemistry and Biochemistry, Georgia Institute of Technology, 901 Atlantic Drive NW, MS&E Bldg., Atlanta, GA 30332-0

Sensing science at ExxonMobil is enabled by a recently assembled Sensor Testing System (**SeTS**) that incorporates a NeSSI^{TM} based multi-analyzer sampling system. The system was designed for flexible gas mixing (currently four gases plus air), rapid sampling to GC, mass spec, spectroscopic, and discrete MEMS or metal oxide based gas sensors. We will report initial experiments with **SeTS**) around a training set composed of benzene, toluene, and xylene gases varied in a concentration range of 1 to 100 ppm resulting in a 100+ member training set. These gases represent an initial set for estimating environmental exposures including hydrocarbons in ambient air, water, and process gas streams. Initial experiments combine conventional GC measurements, mid-infrared FT-IR spectroscopy with waveguide-enhanced sensitivity, and self-assembled monolayer (SAM) coated microcantilevers. A rapid search for selective SAMs is enabled by the (**SeTS** gas mixing and sampling system. Chemometrics-based analysis of multivariate FT-IR data shows substantial promise for accurate predictions at ppm constituent levels and below. We see significant promise for this technology for rapid testing of microanalytics (GC, MS, FT-IR), polymer or metal oxide gas sensors, and associated pre-concentration systems. In future, the flexible NeSSI^{TM}-based design enables the integration of microreactors along with the testing system. Furthermore, rapid transfer from lab testing to plant environments is facilitated by the inherent robustness of NeSSI^{TM}-based sampling systems, and small device footprints

AUTHOR INDEX

AUTHOR INDEX

AUTHOR INDEX

American Institute of Chemical Engineers
3 Park Avenue
New York, NY 10016-5991

ISBN: 978-1-60423-350-6